The ALMA Telescope

ALMA, the Atacama Large Millimeter/submillimeter Array, situated high in the Chilean desert, is the largest ground-based telescope on Earth. This is an insider's account of how this complex mega-project came to fruition from authors with intimate knowledge of its past and present. The separate roots of ALMA in the United States, Europe, and Japan are traced to their merger into an international partnership involving more than 20 countries. The book relates the search for a suitable telescope site, challenges encountered in organization, funding, and construction, and lessons learned along the way. It closes with a review of the most significant results from ALMA, now one of the most productive telescopes in the world. Written for a broad spectrum of readers, including astronomers, engineers, project managers, science historians, government officials, and the general public, the eBook edition is available to download as an Open Access publication on Cambridge Core.

PAUL A. VANDEN BOUT, Ph.D. is a Senior Scientist, Emeritus, at the US National Radio Astronomy Observatory (NRAO), where he served as a director from 1985 to 2002. He was the first director of ALMA and served as the head of the North American ALMA Science Center (NAASC), where he organized the Center in its early years. His career has been almost entirely spent in millimeter astronomy, including pioneering the Millimeter Wave Observatory at the University of Texas' McDonald Observatory. He has participated in the entire US history of events that led to ALMA, and much of that in Europe, Japan, and Chile.

ROBERT L. DICKMAN, Ph.D. is a Scientist, Emeritus, at NRAO. His entire career has been spent in radio and millimeter wavelength astronomy. He was head of the NSF Division of Astronomical Sciences' Radio Astronomy Unit, managing the process of approval for funding the Millimeter Array – ALMA's precursor – and then ALMA itself. As a US Embassy Fellow in Santiago, Chile, he advanced the negotiation to secure the right to build and operate ALMA. After he left NSF, he held senior positions at the NRAO, first in New Mexico and then in Charlottesville, Virginia.

ADELE L. PLUNKETT, Ph.D. is an Associate Scientist at NRAO, working in the NAASC. As a Fulbright Fellow in Chile in 2012, she contributed to the Commissioning and Science Verification Team at ALMA, and in the years that followed, as an European Southern Observatory Fellow, she has spent numerous shifts as Astronomer on Duty at ALMA. She is fluent in Spanish and previously studied Japanese, leading to a serendipitous synergy with the ALMA project.

The ALMA Telescope

The Story of a Science Mega-Project

PAUL A. VANDEN BOUT
National Radio Astronomy Observatory

ROBERT L. DICKMAN
National Radio Astronomy Observatory

ADELE L. PLUNKETT
National Radio Astronomy Observatory

Shaftesbury Road, Cambridge CB2 8EA, United Kingdom

One Liberty Plaza, 20th Floor, New York, NY 10006, USA

477 Williamstown Road, Port Melbourne, VIC 3207, Australia

314–321, 3rd Floor, Plot 3, Splendor Forum, Jasola District Centre, New Delhi – 110025, India

103 Penang Road, #05–06/07, Visioncrest Commercial, Singapore 238467

Cambridge University Press is part of Cambridge University Press & Assessment, a department of the University of Cambridge.

We share the University's mission to contribute to society through the pursuit of education, learning and research at the highest international levels of excellence.

www.cambridge.org
Information on this title: www.cambridge.org/9781009279680

DOI: 10.1017/9781009279727

First published 2023

A catalogue record for this publication is available from the British Library.

Library of Congress Cataloging-in-Publication Data
Names: Vanden Bout, Paul A., 1939– author. | Dickman, R. L. (Robert L.), 1947– author. | Plunkett, Adele L., 1988– author.
Title: The ALMA telescope : the story of a science mega-project / Paul A. Vanden Bout, National Radio Astronomy Observatory, Charlottesville, Virginia (Emeritus), Robert L. Dickman, National Radio Astronomy Observatory, Charlottesville, Virginia (Emeritus), Adele L. Plunkett, National Radio Astronomy Observatory, Charlottesville, Virginia.
Description: Cambridge ; New York, NY : Cambridge University Press, 2023. | Includes bibliographical references and index.
Identifiers: LCCN 2022060274 | ISBN 9781009279680 (paperback) | ISBN 9781009279727 (ebook)
Subjects: LCSH: ALMA (Observatory : Chile) – History. | Atacama Large Millimeter Array (Project) – History. | Millimeter astronomy – History. | Very large array telescopes – Chile – History. | Millimeter astronomy.
Classification: LCC QB479.C5 A46 2023 | DDC 522.2983–dc23/eng/20230317
LC record available at https://lccn.loc.gov/2022060274

ISBN 978-1-009-27968-0 Paperback

This book is dedicated to the memory of
Robert Lamme Brown
Riccardo Giacconi
Richard Edwin Hills
Norio Kaifu
Kwok-Yung Lo
and
Koh-Ichiro Morita
whose vision of ALMA became a reality,
thanks to their ingenuity and diligence.

"There is no richer field of science opened to the exploration of man in search of knowledge than astronomical observation; nor is there ... any duty more impressively incumbent on all human governments than that of furnishing means and facilities and rewards to those who devote the labors of their lives to the indefatigable industry, the unceasing vigilance, and the bright intelligence indispensable to success in these pursuits."

John Quincy Adams, at the dedication of the Cincinnati Observatory, 9 November 1843

Contents

Foreword

As this book goes to press, the Atacama Large Millimeter/submillimeter Array, ALMA, has been making cutting-edge observations of the Cosmos for over 10 years. It is managed jointly by a partnership between North America, Europe, and East Asia. Financial support for operations, as well as for the development construction in Chile, the host country, is also a shared enterprise and comes from the United States National Science Foundation (NSF), the European Southern Observatory (ESO), and the National Institutes of Natural Sciences of Japan (NINS). Canada, South Korea, and Taiwan also contribute funding through these partners. Indeed, over 20 nations are actively involved in ALMA. Worldwide teams of dedicated and persistent scientists and engineers worked together to design, develop, and construct this complex and unique array, beginning in the early 2000s. Sixty-six high-precision radio antennas, 54 of diameter 12 m and 12 of 7 m, each equipped with banks of detectors and supported by complex electronics, are now distributed over the high and dry Chajnantor Plateau that overlooks Chile's Atacama Desert. The technical capabilities and the emergent scientific results have outstripped expectations and are transforming our view of the Universe. Investigations span space and time, from billions of years ago when the first stars and galaxies were forming to the present when innovative and exquisitely detailed dynamical and chemical probes of star and planet formation in our own galaxy are illuminating the origins and evolution of other solar systems. ALMA has enormous promise for continuing such grand discoveries. This book describes how the international partnership came about and how, in spite of the enormous challenges that arose at every phase of this wildly ambitious project, the partners achieved dazzling success. It provides a wealth of detail about ALMA's design, development, and construction along with a sense of the roller coaster of highs and lows that accompanied the resolution of the many difficulties. It should appeal to a wide spectrum of readers.

Although this history views the ALMA project from a very American perspective, the authors bring to the table a hard-to-match acquaintance with many events that shaped the emergence of millimeter/submillimeter wavelength astronomy and the eventual construction of a millimeter/submillimeter observatory hundreds of times more powerful than any that could have been achieved without an international partnership. Their professional involvement extends from the 1970s to the completion of the array and current operations. By the 1980s, Paul A. Vanden Bout, then Director of the US National Radio Astronomy Observatory (NRAO), was a leading figure in efforts to build a large interferometer operating at millimeter wavelengths for the American astronomy community. The concept of the Millimeter Array (MMA) was an important precursor to the US ALMA effort. Robert L. Dickman, as leader of the Radio Facilities Unit of the NSF's Astronomy Division, became responsible for representing and promoting the burgeoning US millimeter wave interferometer program within the agency in the early 1990s. As the MMA plan merged into the international ALMA endeavor, he played a critical role in bringing the fledgling partnership to fruition, particularly in matters that required NSF acquiescence and financial support. Both he and Vanden Bout continued to be intimately involved in the critical decision-making processes for ALMA that followed. Adele L. Plunkett represents the modern face of ALMA. Her Yale Ph.D. dissertation relied partially on measurements from ALMA's first official cycle of observations. She went on to become an ESO Fellow in Chile before taking up her current position at the North American ALMA Science Center (NAASC) in Charlottesville, Virginia. I was amused to learn from this book that, as a beginning graduate student, she was part of an early team making experimental submillimeter observations on a peak near the Chajnantor Plateau.

The authors readily acknowledge that they present a largely American perspective but include relevant details on the evolution of the European and Japanese millimeter wave programs that proceeded independently. Together, these were the foundation for the ALMA partnership. For obvious reasons, the American saga is most in-depth and personal. I will not be surprised to see complementary accounts from Europe and East Asia in the future. That said, I was impressed by the clear intent to provide an accurate and objective recounting here. References to pertinent archives and available documents, some quite obscure, are frequent and verifiable. Moreover, the authors have unearthed numerous written records that might well have eluded others. In addition to the astrophysicists, students, instrument builders, and project managers targeted as readers, historians of science should enjoy the level of detail. Others, like me, may appreciate that very human reactions are on display amid the profusion of well-referenced scientific, technical, and management detail. Many disappointments as well as the joys of accomplishment can be expected along the way to outrageously ambitious instruments such as ALMA. With a view to balancing the American perspective, individual essays – vignettes – with reminiscences by ALMA colleagues from Europe, East Asia, and Chile are included here. Although these are of necessity concise, they demonstrate that all partners experienced a similar range of reactions at one time or another.

I thought that the sometimes-whimsical quotations that precede each chapter also helped make the narrative relatable. For the most part, they set the tone for each chapter rather cleverly. In Chapter 1, Dutch astronomer, Ewine van Dishoek's vignette summarizes nicely how the detection of radiation from the carbon monoxide molecule "opens your eyes." Chapter 7's quotation, "Necessity never makes a good bargain," however, did give me pause. Indeed, the cost of this unique and transformational array made the international partnership necessary. But ALMA is now an undisputed success – surely a good bargain? Some degree of tension among the partners was probably inevitable. By the 1990s, the major research communities centered in the United States, Europe, and Japan had emerged and evolved, following the first tantalizing millimeter wavelength discoveries of interstellar molecules and dust two decades earlier. The years of scientific and technical planning for North America's MMA, not to mention its perennial financial struggles, dominate the beginnings of this ALMA history. However, the scientific potential of submillimeter as well as millimeter detections also inspired ambitious studies and the construction of ever-more-capable telescopes in Japan and Europe and culminated in the Large Millimeter Submillimeter Array (LMSA) and Large Southern Array (LSA) concepts, respectively. Not surprisingly, all three groups had similar and exciting science objectives for their technical visions of a powerful array of many large-diameter millimeter/submillimeter antennas. These are detailed in the initial ALMA proposal text presented here. Many readers may enjoy, as I did, the graphic account of the driving questions provided in ESO astronomer Peter Shaver's succinct vignette. The technical requirements for instruments that matched these objectives were staggering in their complexity and similar for each group. The precision of the surfaces of the many large antennas would have to be a factor of at least a hundred times greater than had yet been achieved for smaller telescopes. Each would be equipped with banks of detector/receivers exquisitely sensitive to the various selected detection bands in the millimeter/submillimeter window and would rely on state-of-the-art electronics. The inherent difficulties in constructing such a unique and complex instrument would be exacerbated by the need for a location on a high, dry site, eventually identified as Chajnantor. Although they came to their decisions independently, Europe, Japan, and North America all agreed on these requirements.

However, as this book emphasizes, the partners also brought different cultural expectations and different fiscal patterns to the ALMA project. The effects were already evident at the time of the first agreements between Europe and North America. Japan was not included, despite an established working relationship between the LMSA and MMA teams. In a poignant vignette Masato Ishiguro, then Director of the Japan's National Radio Observatory, vividly expresses his surprise and dismay. The immediate problem was smoothed over, and Japanese representatives continued to participate at all stages of the ALMA project, but later negative decisions by the Japanese government delayed the entrance of Japan into the ALMA partnership for several years. As a result, ALMA stands alone among large international ground-based observatories not only because of its unique science contributions but also because of its unusual organizational structure. Europe and North America are equal partners with significant but lesser participation by East Asia. Until Japan formally joined the collaboration, the equal partnership between Europe and North America required that they reach acceptable compromise, if not consensus, on every aspect of bringing together the LSA and MMA concepts. The negotiations could be difficult and sometimes tortuous, often continuing over meetings and between meetings, as I witnessed while a member of the ALMA Board representing the US astronomy community. The narrative here, along with the vignette from astronomer Pieter van der Kruit, then Chair of ESO Council, reflects frustrations on both sides of the Atlantic with regard to the long-drawn-out decision-making processes. Van der Kruit has suggested that ALMA's organizational structure contributed significantly to the situation. In fact, the merger of the EU and NA projects began collegially and proceeded well for some time under the bipartite structure. Together, the partners faced and overcame a number of significant challenges. These included budget reviews and the completion of the memoranda of understanding that defined every construction task required to construct ALMA, as well as the associated implementation plans. Negotiations with Chile, the host country, especially concerning the Chajnantor site, were concluded successfully, and preparations for Japan's joining the partnership continued cordially. Although there were some initial impediments when it came to the details of distributing the construction tasks among the partner countries, a small international team of ALMA leaders settled these in a timely manner.

The truly "contentious" matters of Chapter 7 were largely confined to decisions regarding the choice of telescopes and vendors for ALMA, the location in Santiago of the ALMA headquarters, the Joint ALMA Office, and whether North America or Europe would employ the necessary Chilean labor force. The authors have not shied away from presenting the scale and scope of the disagreements.

In these exchanges, cultural differences and funding arrangements often played significant roles. Readers can follow in some detail the considerations involved, and gain some insight into whether, overall, the compromises were less than ideal. What is undeniable is that ALMA, conjured from these negotiations, is a unique, complex, and incredibly powerful instrument that is delivering results far beyond the hopes of its most enthusiastic proponents. Moreover, it is an international endeavor in so many ways. In the end, each partner chose an internal vendor so that the fifty-four 12 m diameter antennas encompass 3 different telescope designs, 25 European, 25 North American, and 4 Japanese. Likewise, teams all over the world assembled the components for banks of sensitive detectors for each telescope and for the innovative associated electronics.

Chapter 10 compares expectations with results and presents an optimistic assessment for ALMA's future. To that assessment, I would add the potential for new discoveries based on the combination of array observations with results from large optical/infrared telescopes, both ground- and space-based. Combining ALMA molecular line and NASA's James Webb Space Telescope measurements, for example, could enable unprecedented studies of the earliest stages of galaxy formation and even add to our understanding of the evolution of dark matter in the Universe. ALMA's scientific bonanza is transforming astronomy as a whole.

At this time when international collaborations to construct ever more sophisticated large telescopes are becoming commonplace, it is well worth considering the question of how ALMA became such an astounding success, despite every challenge that arose. Certainly, the endeavor had its fair share of the "barely survived disasters" alluded to in the Preface. Every project of this scope should expect these. In this regard, the "Lessons Learned" section is insightful and should be heeded, especially where the effects of cultural differences and the role of politics in funding decisions are concerned. On the other hand, it is obvious from the very human reactions that permeate this account that everyone involved in ALMA was passionate about creating an instrument that would enhance our scientific view of the Universe. That spirit fueled the determined perseverance of the ALMA partners and enabled the project to survive the most acrimonious negotiating sessions and the numerous almost-disasters. The authors' words encapsulate the foundation of that spirit and of ALMA's spectacular success. Everyone believed – and believed deeply – "it's all about science." That shines through in this account.

Anneila Sargent
Ira S. Bowen Professor of Astronomy Emeritus
California Institute of Technology, Pasadena

Preface

In March 2013, overlooking the oasis village of San Pedro de Atacama in Region II of the Republic of Chile, nestled in the altiplano above the Atacama Desert, the Atacama Large Millimeter/submillimeter Array (ALMA) was entering operation. ALMA is a partnership between North America, Europe, East Asia, and the host country of Chile. It is a truly global radio astronomy project and, arguably, the most complex ground-based astronomical instrument ever built.

By the end of that same year, the array would consist of 66 radio antennas, with 54 of them 12 m in diameter, and the remaining 12 measuring 7 m across. These antennas are electronically linked and able to operate as one instrument, a telescope with the sensitivity of a single antenna almost 90 m across. Arranged in various configurations, the antennas can also operate in a way that sees as much detail as a much larger telescope, up to 16 km across. By moving the antennas so that their largest separation approaches this distance, ALMA can image the radio sky with more than 10 times the detail of the legendary Hubble Space Telescope. ALMA is designed to operate at radio wavelengths ranging from about 0.35 to 7 mm. This is a huge operating range, and to provide full scientific coverage of the sky at these wavelengths, each of the ALMA antennas is equipped with 10 receivers, each cooled to 4 K (−269 C or −452 F) to minimize background noise.

ALMA was envisioned to provide information on the structure and chemical makeup of astronomical objects as close as the Sun and its planets, and as far away as the stars and galaxies of the early Universe. Its capabilities would reveal never-before-seen details in the formation process of nearby stars and their planetary systems, and the birth of galaxies and their coalescence into growing clusters. In the 10 years after its inauguration, ALMA would anchor a coordinated, worldwide collaboration of radio telescopes that would image the black holes at the center of the galaxy M87 and our own Milky Way. Most exciting of all are the discoveries that will be made by astronomers not yet born, discoveries that cannot as yet be fathomed, but which will certainly need the unparalleled capabilities of the array. ALMA was designed at the outset for decades of operation, built to be improved as technology progresses and the questions deepen.

This book is the story of the 40 years of ALMA, from the death in 1982 of the NRAO's proposal to the NSF for a 25 m radio telescope, to the inauguration of ALMA in 2013, and first decade of operation. It is a complex story. The participants in the ALMA project melded their initial technical visions of how to answer the biggest questions of modern astronomy in a drawn-out ballet of discussions, studies, and negotiations that ended up with a global partnership involving more than 20 nations. Management of the project was challenged by profound cultural differences and the fact that no partner had a majority share, slowing the process for making decisions, but ultimately enabling more than any single partner could have accomplished alone. A near-death crisis occurred when it was found that the initial costs were significantly underestimated. It is a story with more than one lucky break and also a story that ends in success. Today, ALMA is one of the top producers of scientific results in the world.

The authors observed this history first-hand. Paul Vanden Bout was NRAO Director from 1985 to 2003 and oversaw the creation of the proposal to build the Millimeter Array (MMA), the US antecedent of what became ALMA. Vanden Bout served as ALMA Director (interim) between 2002 and 2003 and as the head of the NAASC within NRAO from 2005 to 2007. Bob Dickman began as a program officer within the NSF Division of Astronomical Sciences (AST) in late 1991, with responsibility for the Arecibo Radio Telescope. He became Coordinator of the Radio Facilities Unit, newly formed within AST in 1993, and was given responsibility for advocating within NSF for the ALMA project on behalf of the US astronomy community, and for securing all the approvals required over many years to construct the instrument. He was also a member of the team that drafted and negotiated the first official ALMA agreement with NSF's international partners, and he served on the ALMA Board. Adele Plunkett was Fellow of the ESO stationed at the ALMA headquarters in Santiago, Chile, from 2015 to 2018. She currently works in the NAASC as an NRAO staff scientist.

The decades-long span covered by this book was a period of enormous progress for astronomy, and by the turn of the twenty-first century, millimeter wave astronomy had matured as an observational science. The basic skeleton of the discipline had been fully defined, and much of the low-hanging fruit of discovery had been picked. It was also becoming clear that, in order to be scientifically worth building, the next millimeter wave radio telescope would have to be vastly more capable than any of its predecessors. As a consequence, it would also be correspondingly much more complex and expensive. Probably more than anything else, this single fact drove NSF's insistence, when funding began to flow, that ALMA be an international partnership, and that its construction be subject to far more structured project management and cost accounting processes than had previously been the case at the NSF. Both of these new requirements were major challenges for the ALMA project, and both are now part of the funding culture for US science projects.

We wrote this book to convey to the generation of astronomers now using ALMA how long and difficult it is to start and bring to completion such a project, and the role chance and circumstance appear to play in the course of events. We hope that the audience for the book will prove to be larger than the research astronomers and physicists who work in our field and will include, perhaps, students of project management, science historians, managers in the federal government and at private foundations, and even members of the general public. In establishing the historical record, we have included a level of detail that may go beyond the interest of some of our readers. We beg their indulgence. We believe it would be a mistake to assume that such a diverse potential audience has a background in astronomical science. As a result, where necessary we provide the scientific background required to understand not only the decisions that led to the ALMA instrument but also the increasing excitement generated by the instrument's potential as its design took shape and matured. The subject of radio astronomy is introduced in an appendix written for the lay reader.

Ours is necessarily an American perspective on this history. We have written first of all about what we observed ourselves. The voluminous material available in the NRAO Historical Archives has been invaluable in authenticating our recollections, as has material from numerous other sources. We have placed citations to sources of the material we used in notes at the end of each chapter. Our book could not have been written without the help of professional archivists Ellen Bouton and Heather Cole, who guided us through the vast trove of records in the Archives to relevant material, added items to the Archives that we had located elsewhere, and now maintain an index of the links to the digital items we cite, ensuring that the links will remain functional. Lance Utley and Kristy Davis, NRAO librarians, were most helpful in providing copies of obscure articles in the scientific literature. We are beyond grateful to all of them for their support of our project.

We are also grateful for the invaluable help of Pat Donahoe, a longtime employee of Associated Universities, Inc. (AUI), the science management corporation that operates NRAO for NSF, in assembling and organizing historical material related to the formation of the ALMA partnership; accounts of the subsequent meetings of the partners in committees, telecons, and board meetings; negotiations with the Republic of Chile for the ALMA site; and the procurement of the ALMA antennas. The ALMA Project was more concerned with establishing a record of accomplishment than with organizing the documentation of that record. Donahoe has done us and the Project an invaluable service by sorting and organizing this material. He was uniquely qualified for the task, having served as secretary to the ALMA Board for many years.

We are also indebted to our colleagues and supporters at the NSF who have watched over NRAO for more than 70 years, and whose support, expertise, and insights helped make ALMA a reality: Neal Lane, former NSF Director, who approved the funds for MMA design and development and insisted that the project have international support, graciously granted us an interview to go over the events of that time; Bill Harris, Bob Eisenstein, and Mike Turner, all former NSF Assistant Directors for the NSF Directorate for Mathematics and Physical Sciences, who played key roles in supporting ALMA within NSF and on ALMA oversight bodies; Vernon Pankonin, former AST Program Officer who advanced the cause of the MMA at NSF during and following the Barrett Report, clarified in an interview his activities on behalf of the US millimeter astronomy community during that period; and Hugh van Horn, former AST Director, was helpful in confirming actions at NSF to promote the MMA.

Thanks are due to Anneila Sargent, California Institute of Technology, for providing historical data for the Owens Valley Radio Observatory and the Combined Array for Research in Millimeter-wave Astronomy, and for her excellent Foreword to this book. We are also grateful to Bob Wilson for his account of the discovery of interstellar carbon monoxide and the history of the Bell Telephone Laboratory 7 m Telescope.

We wish to acknowledge the help of colleagues at NRAO past and present, scientists, engineers, and business managers whose expertise helped lay the path for ALMA: Fred Lo, who as NRAO Director guided the North American partnership through the difficult times of antenna procurement, rebaselining the ALMA budget, and the long road that led to East Asia as a third ALMA partner; Frazer Owen, the NRAO scientist who suggested building a "millimeter Very Large Array," and who more than anyone else developed and promoted the MMA concept; Peter Napier, NRAO staff scientist, who served on the committees that developed the specifications for the ALMA 12 m diameter antennas and later on the antenna contract selection committee for NRAO; Bill Porter, NRAO business manager for ALMA, who with Pat Donahoe clarified for us the antenna procurement process; Al Wootten, North American ALMA Project Scientist, whose minutes of the ALMA Science Advisory Committee were most useful; Tony Beasley, current NRAO Director and former ALMA Project Manager for insight into ALMA's biggest crisis and his experience in ALMA management, and for his encouragement of the writing of this book and support of the Archives; Mark Gordon for supplying the details of his study on the feasibility of building and operating the MMA in Chile; Art Symmes, for explaining the process of antenna delivery, assembly, and testing; Harvey Liszt for clarifying the issues surrounding radio frequency interference and protection at the ALMA site; Jeff Mangum for explaining the process of ALMA antenna prototype testing; Nick Emerson, former NRAO engineer, for providing details of the process of ALMA antenna acceptance and commissioning; Jody Bolyard, NRAO Safety Officer, for an explanation of the ALMA safety program; Phil Jewell, NRAO Deputy Director, and Mark McKinnon, former NRAO ALMA Program Manager, for a multitude of facts important to our narrative; and Jeff Hellerman and Bill Saxton for providing the diagrams in the appendix on radio astronomy.

We are grateful to our colleagues at AUI, past and present, who advocated for ALMA: Martha Haynes, AUI Interim President at the time of AUI's certification to do business in Chile; Adam Cohen, current AUI President, for releasing AUI documents to the Archives; Eduardo Hardy for help he gave us in establishing the steps that were made in obtaining use of the ALMA site; and Paulina Bocaz and Rodrigo Brito for supplying us with detailed information on ALMA as it exists today.

We acknowledge our colleagues in the international astronomical community, whose intellects helped foreshadow ALMA and give it concrete form: Peter Shaver, ESO scientist, who successfully marshalled the effort in Europe to realize Roy Booth's vision of a large millimeter array in the Southern Hemisphere. We deeply appreciate his help in establishing for us the European narrative of ALMA; Masato Ishiguro, scientist at the National Astronomical Observatory of Japan, who was one of the key individuals in the development of the Large Millimeter/Submillimeter Array and the path to Japan's entrance into ALMA. We are most grateful to him for his willingness to share the material he assembled for his own account of millimeter astronomy in Japan, from the Nobeyama Radio Observatory to ALMA; Pierre Cox and Thijs de Graauw, former ALMA Directors, who described the challenges that come with that position; Ewine van Dishoek for documents relevant to the Netherlands joining the MMA; Dennis Downes for material setting out his vision for the LSA; Pierre Encrenaz and Alain Omont for early history of the Institut de Radio Astronomie Millimétrique; Alain Baudry for early history of millimeter interferometry at the Bordeaux Observatory; Richard Kurz for accounts of events in Europe and Chile; Torben Anderson and Stephan Guilloteau for their memories of the meeting where it was agreed to pursue a merger of the MMA and LSA;

Sean Dougherty, ALMA Director at the time of the writing of this book, for his insights into ALMA operations; and Alejandra Voigt, Executive Officer at the Joint ALMA Observatory, for releasing ALMA Board meeting documents that helped establish the course of events.

We wish to thank April Burke and Michael Ledford of Lewis-Burke Associates, the former government relations firm for AUI; Carol McGuire, former long-time US Senate Budget Committee staffer for Senator P.V. Domenici of New Mexico; and Tim Clancy, NSF Congressional Liaison and Public Affairs staff member during Rita Colwell's tenure as NSF Director. All were helpful in providing details of the funding of the MMA/ALMA by the US Congress.

We are grateful for the advice of Ken Kellermann, Ron Ekers, and Miller Goss regarding selection of a publisher. Our special thanks to Vince Higgs, our editor at Cambridge University Press, for advice and guidance, and to Sarah Armstrong for helpful assistance.

In an attempt to broaden the perspective of our book, we have included vignettes written by colleagues in Europe, Japan, and Chile throughout the narrative. We are most grateful to Leo Bronfman, Paulo Cortes, Ewine van Dishoeck, Pat Donahoe, Viviana Guzmán, Masato Ishiguro, Pieter van de Kruit, Angel Otárola, Masao Saito, and Peter Shaver for their contributions.

Several individuals reviewed our draft manuscript, either particular parts or the entire book, and we are most grateful for helpful comments from Ellen Bouton, Butler Burton, John Carpenter, Rebecca Charbonneau, Ineke Dickman, Antonio Hales, Masato Ishiguro, Criss Cloudt, Anneila Sargent, and Peter Shaver. They uncovered numerous mistakes; we take ownership of those that remain.

Finally, we wish to thank Rachel, Ineke, and Patricio for their encouragement and patience throughout the writing of this book and during our decades of involvement with the ALMA project and adventures along the way.

Interstellar Carbon Monoxide

"There are poisons that blind you, and poisons that open your eyes."

August Strindberg, The Ghost Sonata

Celebrating a Toxic Gas

At a resort in the foothills of the Catalina Mountains near Tucson, Arizona, over 225 astronomers from 29 countries gathered on 29 May–5 June 1995 for Symposium No. 170 of the International Astronomical Union, to discuss the topic *CO: Twenty-Five Years of Millimeter-Wave Spectroscopy.*[1] The symposium was organized by the National Radio Astronomy Observatory (NRAO) together with astronomers at the University of Arizona's Steward Observatory. By coincidence, the venue was shared with a convention of the National Rifle Association (NRA). There was confusion over the acronyms on the name tags, the NRA people thinking the astronomers were part of their meeting and vice versa. The gun fans never understood why the astronomers were celebrating the discovery of a gas in outer space, albeit, a gas as lethal as their rifles. Indeed, what was the fascination to astronomers of interstellar carbon monoxide? The explanation begins much earlier.

Interstellar Medium

Astronomers were slow to realize that the space between the stars was not empty. Over the course of the first half of the twentieth century, they came to recognize that a thin haze of gas and microscopic dust particles permeates the space between the stars in the disk of our Milky Way Galaxy. This is the interstellar medium (ISM). The dark voids seen in the constellations of stars were not devoid of stars. Rather, they were vast, intervening clouds of dust, obscuring our view of the stars that lay behind them. The Atacameños, who had long occupied the area in Chile where ALMA would be built, regarded the dark nebulae of the Southern Sky as their constellations,[2] rather than the stellar constellations familiar to us. Light from stars lying behind the dust clouds is both absorbed and, more importantly, scattered into new directions by the dust. The dust clouds are opaque to visible light much the way that clouds of smoke from fires and dust storms on Earth can obscure our vision. Lists[3] have been made by astronomers of those dust clouds that have reasonably well-defined shapes. The Atacameños saw llamas, birds, shepherds, and foxes, to cite a few examples. It is a wonderful and fitting coincidence that ALMA is now focused on these dark nebulae for a good portion of its observing time.

The first molecule to be discovered in interstellar space was methylidyne (CH) in 1938, soon followed by CH$^+$ and cyanide (CN). Technically, these molecules are free radicals, highly reactive bits of molecules that exist only fleetingly in the laboratory. In regions of interstellar space, the density of gas and dust is low, lower than the best laboratory vacuum, sufficiently slowing the reactions that destroy the free radicals to make them observable. They were identified by their narrow absorption lines seen in the spectra of bright stars.[4] It took 12 more years before another interstellar free radical was discovered, the hydroxyl radical, OH, by the detection of its spectral line at radio wavelengths.[5] In 1957, Charles Townes discussed the probabilities that other molecules could exist in the ISM.[6] On his arrival in 1967 at the University of California, Berkeley, Townes began a search for molecules in space, starting with ammonia (NH$_3$). In the fall of 1968, he detected ammonia, and followed that the next year by detecting water (H$_2$O).[7] Interstellar formaldehyde (H$_2$CO) was discovered[8] using the NRAO 140 Foot Telescope in 1969. The next year, interstellar carbon monoxide (CO) was discovered in space.

Discovery of Interstellar Carbon Monoxide

The first millimeter wavelength observations at NRAO occurred at its Green Bank, West Virginia, site. Frank Low had moved to Green Bank from Texas Instruments in Dallas, Texas, where he had built a detector of infrared/millimeter radiation using a germanium chip. But the observations made with a small trial dish quickly showed that a drier site with better atmospheric transparency was needed. In 1962, NRAO requested funding from the NSF for a millimeter telescope of diameter 36 ft. Funding was received that same year and Kitt Peak, Arizona, was chosen as the site. It had acceptable atmospheric quality and a well-developed infrastructure. After a difficult period of construction, the telescope came into operation in 1968.[9]

It was intended for continuum observations, that is, recording the strength of millimeter emission in broad chunks of the spectrum. The telescope was also equipped with a filter bank that allowed for spectral line observations.

The team that first detected interstellar CO was composed of Bob Wilson, Keith Jefferts, and Arno Penzias, all of Bell Telephone Laboratories. Their letter[10] requesting observing time on the 36 Foot Telescope stated as the primary goal the detection of CN. Searches for CO and then hydrogen cyanide (HCN) were next in priority. These choices were based on consideration of a number of candidate molecules in the list of Townes and talking with Pat Thaddeus who pointed out the proximity of CN lines to those of CO. Phil Solomon had suggested looking for CO. He thought CO would be an abundant ISM molecule based on its properties.[11] The proposal letter noted that, "*Although CO has a smaller dipole moment than CN, it is also worth looking for owing to the fact that its dissociation energy is above the Lyman continuum making it potentially much more abundant than the other gases heretofore detected in which this is not the case.*" The proposal received only "Average" and "Fair" ratings but was scheduled for observing time nonetheless. Once at the telescope, the team first chose to look for CO. They pointed the antenna to the Orion Nebula, a bright nebula that was overhead at the time their observing run began. It turned out to be the strongest CO source in the sky.[12] They detected it within seconds.

Prior to the observations of interstellar molecules that the detection of CO spawned, the 36 Foot Telescope, shown in Figure 1.1, was not in high demand. It was used mainly to measure the brightness of planets and other radio sources at millimeter wavelengths. The discovery of CO dramatically changed that, and it became NRAO's most popular telescope for a considerable period of time. Its capability at millimeter wavelengths meant it was ideal for observations of interstellar CO and other molecules with spectral lines in the millimeter band. The 36 Foot (later 12 Meter) Telescope remained a productive facility until it was closed in July 2000 to direct resources to ALMA.

The detection[13] of interstellar CO, shown in Figure 1.2, was a watershed discovery, leading to new areas of research in astronomy and transforming others. It all had to do with the formation of new stars. Astronomers knew a great deal at the time about how stars evolve to end their lives, but little about stellar birth. It was assumed that stars formed by the action of gravity, pulling together a stellar mass of material from surrounding space. Star formation was thought to occur in regions where the ISM was of high density. But the high density also meant those regions were unobservable in visible light, no matter how large the (optical) telescope might be. Any light emitted by a newly formed star would be scattered by the surrounding dust, rendering it invisible to an optical telescope. Fortunately, the dust in these regions does not scatter the much longer wavelengths observed by radio telescopes. And the CO in these regions turned out to

Figure 1.1 The NRAO 36 Foot Telescope in its fabric-covered astrodome on Kitt Peak, Arizona. Credit: NRAO/AUI/NSF, CC BY 3.0.

produce a remarkably strong radio signal. For the first time, astronomers could measure the physical properties of star-forming regions of space without their view being obscured by the interstellar dust clouds, allowing them to formulate theories of star formation based on observational data. Fifty years later, ALMA would continue to explore the potential of CO for star formation studies, not only in the Milky Way but in distant galaxies in the Universe.

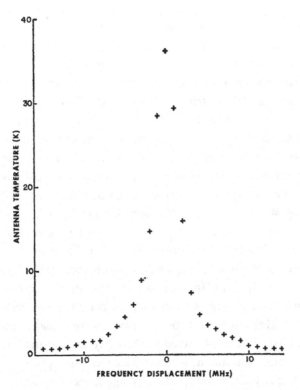

Figure 1.2 The CO signal in the Orion Nebula as first detected. The plot shows the signal strongest at the exact frequency expected for CO in this source, falling dramatically within about ±10 MHz from that frequency. Credit: Wilson, Jefferts, and Penzias (1970); ©AAS, reproduced by permission.

ALMA and the CO Molecule

As an astrochemist, I became involved in the planning of what would become ALMA in the early 1990s, both on the US and European sides. It was clear from the beginning that ALMA would be the "astrochemistry machine" of the future, observing both simple and complex molecules. At that time, CO was already commonly and easily observed in virtually every astronomical source, from planets in our own solar system to star-forming clouds and nearby galaxies, with the high-redshift universe just being opened up in CO.

Although CO is a simple diatomic molecule, its formation in space is very different from that in a laboratory on Earth. The actual chemistry of interstellar CO was elucidated in the 1970s–1980s thanks to close collaboration between astronomers and chemical physicists. The molecule is formed through a series of ion–molecule reactions starting with C^+ and OH, but once formed it is hard to destroy because of its very strong triple bond: its binding energy of 11 eV is more than double that of other molecules. Only hard UV photons and reactions with He^+ can break CO back into atoms, but even UV destruction is not very effective because of CO self-shielding – a process studied in detail by John Black and myself. Since CO locks up the bulk of the available carbon, the CO abundance is rather constant, so CO is often used to trace the mass of molecular gas because the dominant interstellar molecule, H_2, is invisible in cold gas.

One of the three key science goals in the design of ALMA was to study disks around young stars, specifically "*to image the gas kinematics in solar-mass protoplanetary disks at a distance of 150 pc, ..., enabling one to study the physical, chemical and magnetic field structure of the disk and to detect the tidal gaps by planets undergoing formation*". ALMA has certainly more than lived up to this challenge. Not only has it imaged exquisite (sub)structures in the dust continuum indicative of planet formation "in action" but it has also revealed gaps and cavities using CO. Those gas cavities are actually smaller than those of the dust, just as predicted by hydrodynamical models. To draw such conclusions and measure gas surface density profiles, the sensitivity of ALMA is key since one needs to image lines of optically thin CO isotopologs, like ^{13}CO, $C^{18}O$, and in some cases even $^{13}C^{17}O$! Vertical structure also plays a role: in fact, subtle isotope selective processes in the disk surface layers can find their ways into ices and minerals that become the building blocks of planets and that can explain isotope anomalies found in solar system meteorites.

Now that ALMA has been operational for more than a decade, CO continues to surprise us. Our surveys of hundreds of disks have shown surprisingly weak CO emission in the bulk of them, by more than an order of magnitude compared with pre-ALMA models. This is not just due to CO being frozen out onto dust grains in the cold outer part of the disk forming an icy mantle. It seems that under cold high-density conditions, CO is being transformed into other molecules like CO_2, CH_3OH, and hydrocarbons. Moreover, grains grow from interstellar (sub)micron size to pebbles and rocks in disks, which can lock up CO and other ices thereby making them invisible. Thus, even a simple molecule like CO can teach us a lot about the first steps in making planets!

Ewine F. van Dishoeck
Leiden Observatory
The Netherlands

Early Development of Interstellar Molecular Astronomy

The detection of CO, and subsequent detections of many other interstellar molecules, spawned vigorous programs in interstellar molecular spectroscopy at radio telescopes equipped to make millimeter wavelength observations, both in the United States and abroad. By 2022, 267 molecular species were known to exist in outer space.[14] Thirty-six of these have 10 or more atoms (three of these are fullerenes, with 60 or more atoms), complex enough to stimulate speculation about life in space. Many of these molecules were detected with the NRAO 36 Foot Telescope which became the 12 Meter Telescope after a replacement of the reflecting surface in 1983. In Appendix B, we give brief descriptions of the pioneering single-dish millimeter facilities that were in operation in the 1970s, whose success led to the proposal by NRAO to build a 25 m diameter telescope. We also present the subsequent telescopes that were built until recently.

The community of radio astronomers in the United States working at millimeter wavelengths had always felt that they owned the field. After all, they had discovered virtually all the interstellar molecules known prior to 1980, and their pioneering facilities had produced a wealth of information on molecular line sources, the distribution of molecular gas in the Galaxy, and star formation in giant molecular clouds. But the advent abroad of the Institut Radio Astronomie Millimétrique (IRAM) 30 Meter Telescope and the Nobeyama Radio Observatory (NRO) 45 Meter Telescope threatened their leadership in the field. Furthermore, the pioneering facilities had reached their limit for many of the scientific questions being asked. For these reasons, it was agreed that the United States needed a large millimeter telescope. Beginning in 1974, NRAO began studying possible replacements for the 36 Foot Telescope. By July 1977, a conceptual design was complete. The telescope would have a diameter of 25 m, be contained in a protective astrodome with a door that could be opened to the sky, and be located on Maunakea, on the Big Island of Hawaii. A formal proposal was submitted to the NSF for funding.

The proposal received outstanding reviews at the NSF and funding seemed likely. So likely, in fact, that the Decadal Review of Astronomy and Astrophysics for the 1980s (the Field Report) considered it to be a project underway and not needing endorsement. But as it turned out, the future of US millimeter wavelength astronomy would need to lie in some other new facility. That is the subject of the next chapter.

Notes

1 The symposium proceedings were edited by Latter et al. (1997).

2 Technically, the Atacameño "constellations" are *asterisms*, as they are not part of the official 89 constellations recognized by the International Astronomical Union.

3 The dark nebulae targeted by observers for CO observations, for example, L134, L1551, B2, etc., came from the lists of Bok and Reilly (1947) and Lynds (1962).

4 Gerhard Herzberg (1988) summarized the discovery of the first interstellar molecular lines in historical remarks submitted to the *Journal of the Royal Astronomical Society of Canada*. Herzberg received the 1971 Nobel Prize in Chemistry for his contributions to our understanding of the quantum structure of molecules.

5 Sander ("Sandy") Weinreb built a new type of spectrometer for analyzing radio signals, an auto-correlator, as his MIT dissertation project. Using the autocorrelator, he discovered interstellar OH (Weinreb et al., 1963). His invention became the standard spectrometer at many radio telescopes. For many years, Weinreb was the lead electronics engineer at NRAO, making many state-of-the-art contributions to radio astronomy technology.

6 Charles Townes (1957) gave a talk at an IAU symposium in which he presented a list of molecules that he thought were likely to exist in the ISM.

7 Townes (2006) reports the history of the discovery of interstellar water and ammonia.

8 Formaldehyde was the first interstellar molecule to be discovered with an NRAO telescope (Snyder, 1969). It was seen in absorption against numerous sources of continuum radiation. That is, the radiation from a distant source would be absorbed by the formaldehyde, creating spectral lines. Later observers realized that H_2CO lines could be seen against the cosmic background radiation. Many other discoveries of ISM molecules at NRAO would follow, largely with the 36 Foot Telescope.

9 For a history of the 36 Foot/12 Meter Telescope, see Gordon (2005).

10 The letter from A. Penzias to William E. Howard, 27 February 1969, and the reply from D. Heeschen to A. Penzias, 18 March 1969 can be found at NAA-NRAO, Tucson Operations, 36 Foot Telescope, Box 4. https://science.nrao.edu/about/publications/alma

11 Solomon and Wickramasinghe (1969) had shown that in regions with a density larger than 100 atoms per cm^3, the gas, overwhelmingly made up of hydrogen atoms (interstellar H had been discovered in 1951 by Ewen and Purcell), would become molecular hydrogen (H_2). After hydrogen, carbon, nitrogen, and oxygen are the most abundant elements in the Universe. The strong bond between C and O in CO meant that once formed it was not easily broken apart and was, potentially, an abundant molecule in the ISM. In fact, we now know it is the most abundant interstellar molecule after H_2.

12 Wilson has published two accounts of the discovery of CO (Wilson, 2008, 2015); the 2008 account is longer.

13 The spectrum is taken from the discovery paper (Wilson, Jefferts, and Penzias, 1970).

14 For a complete list as of 2021, see McGuire (2022).

2

What Now?

"Failure is simply the opportunity to begin again, this time more intelligently."

Henry Ford

Demise of the 25 Meter Project

US National Science Foundation, 1800 G Street NW, Washington, DC, April 6, 1982: A vote of the Advisory Committee for Astronomical Sciences (ACAST) on a motion to endorse the construction of the 25 m diameter millimeter wavelength telescope proposed by NRAO was defeated, with three in favor, seven opposed, and one abstention.[1] The proposal was essentially dead and shortly thereafter NRAO withdrew it. In reality, the project had all but died months before. It had become a victim of inflation, political necessity, and a growing unease in the US millimeter astronomy community about what it really wanted next in the way of a new telescope.[2]

Support for ground-based radio astronomy research in the United States – including operational support for the NRAO and funding for new telescopes – comes, with few exceptions, through NSF. Since being proposed to NSF in 1977, the 25 Meter proposal had experienced a number of existential crises but had nevertheless remained on NSF's radar screen for potential future projects. By late 1981, the 25 Meter had simply spent too much time waiting in the wings while the state of millimeter astronomy and the priorities of the US astronomy community (and NRAO) had continued to evolve. The ACAST vote, while only formally advisory to NSF, was proof of the fact that the consensus of the US astronomy community for the project had eroded beyond the point of viability. The decision of the incoming Reagan administration to put a hold on all new project starts in the Federal budget for the fiscal year 1982, made in the face of an alarming inflation rate of nearly 15 percent, was the final blow.

The loss of the 25 Meter was regrettable. From the start, the proposed telescope had been intended to retain and extend the leadership of the United States in millimeter wavelength astronomy, a field of research the country had pioneered in the late 1960s. By the late 1970s, that leadership was being challenged by the start of construction in Europe of the IRAM 30 Meter Telescope and the 45 Meter Telescope in Japan.

But while Europe did take over leadership in some areas of millimeter wave astronomy, the advance of neither the science nor its supporting technologies slowed in the United States during the 1980s. The field, which became established around 1970, was barely a decade old, and there was much in the way of low-hanging scientific fruit that needed only imagination and opportunity to fall into the hands of observers. In the absence of a large, new US antenna, American millimeter wave astronomers continued to develop innovative research programs that were structured around available domestic millimeter wavelength telescopes. At the same time, engineering research in the United States was leading the way to progressively more sensitive and robust millimeter and submillimeter wave receivers. Even more important, during the late 1970s and early 1980s, groups at the Hat Creek Radio Observatory (HCRO) of the University of California, Berkeley, and at the California Institute of Technology (Caltech) Owens Valley Radio Observatory (OVRO) had begun construction on the first generation of millimeter wave radio interferometers.

Although by present standards those instruments' early sensitivities and imaging capabilities would be modest, they were paving an important groundbreaking path. Interferometers have an intrinsic advantage relative to single-antenna radio telescopes. The imaging detail possible with an interferometer can be enhanced by simply adding antennas separated by progressively greater distances and extending the computational capacity of the signal correlator. This scalability offered millimeter astronomers an alluring pathway for planning instruments vastly more sensitive than the existing millimeter wave radio telescopes, and with the potential to make images of the sky even more detailed than those of the still-to-be launched, Hubble Space Telescope. The outcomes of these projects are described in the sections to come.

With the collapse of NRAO's 25 Meter project in April 1982, one strategic factor stood out clearly: US leadership in millimeter wave radio astronomy demanded a large, new world-class instrument for US radio astronomers. Under the aegis of NSF's Division of Astronomical Sciences, an effort to refocus scientific and political support for such an instrument began almost immediately. The as yet incomplete California interferometers would be crucial guide posts in this process, and their lineal descendant, the Millimeter Array (MMA), only dimly grasped at this time, would evolve and take the shape over 30 years to become part of the largest ground-based radio telescope yet: the Atacama Large Millimeter/submillimeter Array (ALMA).

Community Action

The US millimeter astronomy community was keenly disappointed at the loss of the 25 Meter. They focused the blame on NRAO even though forces beyond NRAO's control – inflation-driven escalating costs and federal budget constraints – were more responsible. The community's first constructive response was to hold a meeting at Bell Labs Crawford Hill to discuss the future. Representatives from NRAO were not invited. The organizers of the meeting were Bob Wilson (Bell Labs), Phil Solomon (SUNY Stony Brook), and Lew Snyder (U. Illinois, Urbana). Fifteen accepted the invitation to attend.[3] The letter of invitation[4] asked what new instrument would be best in the circumstances, an "mm VLA" or a cheap 25 m class telescope? By "mm VLA" the organizers meant an array like the Very Large Array operated by NRAO but capable of millimeter wavelength observations.

Discussion quickly led to a consensus that an "mm VLA" would be the instrument for the future. The group composed a letter[5] outlining the case for a millimeter array to Pat Bautz and her superior Francis Johnson, Director of NSF's AST and NSF Assistant Director for the Directorate of Astronomy, Atmospheres, Earth, and Oceans, respectively. Addressing the letter to Bautz's superior was done to ensure a response. The science goals of the array were star and planetary formation; structure, evolution, and dynamics of galaxies; and cosmology. The letter was sent on 29 October 1982, and Bautz replied to it on 23 November 1982.[6] The group's lack of confidence in Bautz, shown by addressing the letter to Johnson as well, was unwarranted. Before receiving the letter, she had already appointed a committee to advise on the future needs of millimeter and submillimeter wavelength astronomy. It was to hold its first meeting on 3 December 1982. The committee came to include four who were present at the Bell Labs meeting, in particular, Alan Barrett of MIT, who was to chair the NSF committee.

The Barrett Committee was technically a subcommittee of the ACAST, the very committee that had refused to endorse the 25 Meter Telescope. There was clearly an awareness at NSF of the need to address the future of US millimeter astronomy in the wake of the 25 Meter's demise. The Barrett Committee was charged with examining the following questions: What are the emerging emphases in millimeter/submillimeter science? Are the millimeter/submillimeter facilities in the United States and abroad adequate to address the science? Are these facilities accessible to US scientists? What new facilities and instrumentation are required, and what are the relative priorities for implementation? A report was requested in time for the April 1983 meeting of the ACAST.

The principal speakers at the Barrett Committee's first meeting were: Jack Welch who discussed the existing millimeter interferometers at the Hat Creek and Owens Valley observatories; Ron Ekers, who presented a new look at millimeter wavelength possibilities; Frazer Owen, who presented a concept for an "mm VLA"; Mark Gordon, who showed a different concept of a 25 Meter Telescope; Tom Phillips, who presented a concept for a submillimeter telescope for Maunakea: Charles Lada, who showed plans for a joint U. Arizona/Max Planck Institut für Radioastronomie (MPIfR) telescope; and Dennis Downes, who reviewed the IRAM facilities. A concept for a giant fixed reflector operating at millimeter wavelengths, similar in design to the Arecibo Telescope, was proposed by Frank Drake.

The Barrett Committee met on two more occasions. A meeting at Bell Labs in February 1983 discussed the scientific justification for a large millimeter array in a series of presentations. This meeting was widely attended by the community and helped form the basis of a broad consensus on what should be done. The third and final meeting of the Committee was held in Chicago, in April 1983, to write their report.

The Barrett Report

The Barrett Committee met their deadline, finishing a report[7] in time for submission to the ACAST at their April 1983 meeting. With respect to the questions posed in the charge to the committee, they found the following:

- The fact that millimeter/submillimeter radiation penetrates the densest interstellar clouds of matter, in contrast to other wavelengths, opens up entirely new opportunities for the study of star formation, galactic structure, and the evolution of galaxies.
- To make advances in these areas requires an instrument with 1 arcsecond resolution or better at the frequency of the CO (1-0) transition, a total collecting area of 1,000–2,000 m^2, and useable spectral coverage to 1 mm wavelength. No existing instrument met these requirements.
- Foreign instruments under construction have not concluded their guest observing policies but, in any event, would not be able to serve the entire US community.

- The next steps should be: a design study of a facility that meets the above requirements, defining the array, site, and costs; the construction and provision for operation of a 10 m diameter class submillimeter telescope on a dry site; and the provision of support for research and development at millimeter/submillimeter wavelengths: support for scientists, upgrading existing facilities, and the development of new technology.

The report then discussed potential research that could be addressed by the recommended new facilities. As we will see, ALMA's achievements have far exceeded the highest hopes and wildest dreams of the Barrett Committee. ACAST endorsed the report, transmitting it to the NSF and encouraging the AST to support its recommendations. NSF welcomed and embraced the recommendations to the extent that all were fulfilled. Superficially, this seems remarkable. In fact, AST had taken pains during the course of the Committee's deliberations to encourage a robust future for US millimeter astronomy and for a national millimeter array. Vernon Pankonin, who was the liaison between the NSF and the Committee, deserves credit for his encouragement of the Foundation and coordination with the community in addressing a matter of great urgency to US millimeter astronomy. It was a watershed moment for US millimeter/submillimeter astronomy that led to the MMA being proposed as a new national millimeter wavelength facility, one that eventually became part of the international partnership called ALMA. It also was an outstanding example of how a unified scientific community and a supportive funding agency could work together.

The report received comment in the scientific press.[8] An article in *Nature* by Peter David noted that the recommendation of a millimeter array was timely, coming just as President Reagan's budget request to Congress was released, which included a 24 percent increase for NSF astronomy for the fiscal year 1984. Mitch Waldrop opined in a *Science* article that a millimeter array might not be built until 1990 given that it was behind the Very Long Baseline Array (VLBA) and other projects at NSF. It did, in fact, take considerable time for the recommendation of a national millimeter array to be realized; ALMA was not inaugurated until 2013.

Impact of the Barrett Report

AST took the advice in the Barrett Report seriously. Although not stated in the report, there was an assumption, supported by NSF, that a national millimeter array would be developed by NRAO. But NSF did not leave it at that. The millimeter community received generous support for their research and development of new instrumentation, in particular, for the construction and operation of millimeter interferometers as well as the establishment of a submillimeter observatory. When the Barrett Report was issued, there were two millimeter interferometers[9] under development in the United States, largely funded by the NSF, one at HCRO and another at OVRO, both with three antennas at that time. Both observatories enjoyed NSF support that allowed pioneering developments in millimeter astronomy, while being managed by universities.

University of California, Berkeley, and HCRO – The HCRO was founded in 1958 to study interstellar atomic hydrogen. This work was done with an 85 ft telescope, built in 1962. Work to enable millimeter wavelength observations at HCRO began in the early 1970s with an interferometer of two antennas. The pioneering work of Jack Welch produced the first scientific results obtained with a millimeter interferometer and those will be discussed in Chapter 3.

HCRO Interferometer and BIMA – Work had begun at HCRO with a two-element interferometer consisting of a 3 m and a 6.1 m antenna operating at 22 GHz, and observing at 3 mm began in 1978. A third antenna, of a newer design and diameter 6.1 m, became operational in 1985. In 1988, an agreement was concluded between U.C. Berkeley, U. Illinois, and U. Maryland to form the Berkeley–Illinois–Maryland Association (BIMA). This provided an infusion of funds for the construction of nine new antennas. Funding also came from an insurance settlement following the loss of the HCRO 85 foot antenna in a storm. After the original two antennas were scrapped, BIMA was an array of ten 6.1 m antennas. An extension of baselines enabled sub-arcsecond resolution observations. New receivers incorporating superconductor-insulator-superconductor (SIS) receivers supported observations at 1 mm wavelength. Jack Welch has described BIMA's instrumentation,[10] and Dick Plambeck has given an account of BIMA's scientific results.[11] Hundreds of publications reported the results of BIMA observations, from the mapping of molecular species in the circumstellar envelope of the evolved star IRC+10216 to the study of solar flares, the first detection of interstellar acetic acid by Lew Snyder, and the CO Survey Of Normal Galaxies (SONG), to highlight only a few. Significant technical advances came out of HCRO besides the development of millimeter wavelength interferometry. For example, the wideband Gunn oscillators developed by John Carlstrom as a graduate student are found at radio observatories worldwide. The BIMA array is shown in Figure 2.1.

In 2004, BIMA was decommissioned and the nine newest antennas were moved to a site above OVRO to join with the six antennas of the OVRO millimeter array to form the Combined Array for Research in Millimeter-wave Astronomy (CARMA).

Figure 2.1 BIMA in its compact configuration. Jack Welch in the foreground. Credit: Plambeck (2006); ©ASP, reproduced by permission.

California Institute of Technology, and OVRO – The first radio telescope, a 32 ft diameter dish, was installed at the OVRO in 1958 under the leadership of John Bolton and Gordon Stanley. It was quickly followed by two 90 ft diameter antennas. Ten years later, a 130 ft diameter telescope was built. These telescopes were used to study a wide variety of phenomena at centimeter wavelengths. The start of millimeter wavelength astronomy at OVRO began with the development and construction of innovative 10.4 m diameter telescopes for a millimeter array. OVRO is known for its technical contributions to radio astronomy.

OVRO Millimeter Array – Beginning in the late 1970s, Caltech astronomers began a program to build a millimeter wavelength array at OVRO. It was centered on 10.4 m diameter antennas built using an innovative design and construction technique[12] developed by Robert Leighton starting in 1974. By 1978, the OVRO array had three 10.4 m diameter antennas although they were not operational until later. Leighton's reflectors were an assembly of hexagonal aluminum–honeycomb sandwich tiles mounted to a carefully designed backup (support) structure. The surface of an assembly was precision cut by a rotating blade that traveled in the radial direction on a precision arm. The assembly rotated underneath the arm on the precision air bearing that had been used to grind

Figure 2.2 The OVRO Millimeter Array. Courtesy of Anneila Sargent, reproduced by permission.

the mirror for the 200 Inch Telescope on Mt. Palomar. After the cutting operation, aluminum sheeting was secured to the surface of each shaped honeycomb tile. The assembly could be taken apart, shipped, and reassembled with high accuracy. The accuracy of the antenna surface allowed for efficient observing to wavelengths as short as 1 mm in the right conditions. Over time, three more of these antennas were added, the last in 1996. During and after the development of the array, until the formation of the CARMA, an extensive program of research was conducted with the OVRO array that was heavily focused on CO in external galaxies,[13] in particular, merging galaxies like Arp 220, Galactic star formation, and emission surrounding young stars[14] such as HL Tauri, which could indicate the presence of a protoplanetary disk. The six antennas comprising the OVRO Millimeter Array are shown in Figure 2.2. Figure 2.3 records the group that took an impromptu hike in the Inyo Mountains to view potential sites for future millimeter arrays. The hike occurred during a conference, 14–16 October 1994, celebrating the scientific results coming out of the OVRO Millimeter Array. A site nearby would later host CARMA.

Combined Array for Research in Millimeter-wave Astronomy – The speed with which an interferometer can image the sky depends on the number of elements among other factors. Combining the OVRO Millimeter Array of six antennas with BIMA's nine antennas would make for a much more powerful array. That fact, plus growing concern over the cost of supporting two independent university-based millimeter arrays, led the NSF to urge the two groups to find a way to merge their efforts. The result was CARMA,[15] a 15-element array, shown in Figure 2.4. The site

Figure 2.3 The "search party" at a place near the Bristle Cone Pines. Left to right: John Carlstrom, Steve Scott, Harry Hardebeck, Anneila Sargent, Paul Vanden Bout, Pat Thaddeus, Nick Scoville, and Phil Solomon. Courtesy of Dave Woody, reproduced by permission.

Figure 2.4 CARMA: six OVRO 10.4 m antennas on the left and nine BIMA 10 m antennas to the right. Courtesy of Anneila Sargent, reproduced by permission.

was above the Owens Valley on Cedar Flat at 2,200 m elevation in the Inyo National Forest. Cedar Flat allowed for up to 2 km separation of antennas, corresponding to 0.13 arcsecond resolution at 1 mm wavelength. CARMA was inaugurated on 5 May 2006. Starting in 2008, CARMA saw the addition of eight 3.5 m diameter antennas from the University of Chicago – John Carlstrom's Sunyaev-Zel'dovich Array. The project manager for CARMA construction was Tony Beasley, who was able to persuade the US Forest Service to grant permission for the construction in an environmentally sensitive area only 15 km from the Methuselah Bristle Cone Pines Grove. CARMA provided a powerful tool for US millimeter astronomers as a counterpart to the IRAM interferometer on the Plateau de Bure. Thirty percent of the CARMA observing time was reserved for observers in the community at large. The large number of users, the power of the array, and excellent observing conditions resulted in an impressive scientific record, with topics covering the Solar System, Galaxy, and distant Universe. CARMA operated from 2005 until 2015, when the NSF stopped supporting all university-based millimeter radio observatories. Today Cedar Flat displays not a hint of CARMA. Six of the 6.1 m BIMA telescopes were sold to the University of Arizona to be used as ground stations in Arizona and Colorado. The rest of the telescopes were moved back to the Owens Valley for storage. Cedar Flat has been totally restored, down to the wild flowers that were catalogued before construction.

Submillimeter Array – A project of the Smithsonian Astrophysical Observatory (SAO) was contemporaneous with the development of the above NSF-supported projects. The Submillimeter Array (SMA) was the vision of SAO director Irwin Shapiro for the participation of the Harvard Center for Astrophysics (CfA) in millimeter/submillimeter astronomy. (The CfA is an umbrella organization that encompasses SAO and the Harvard College Observatory.) Shapiro initiated the SMA in 1983. Construction took place on a site in the "saddle" of Maunakea, which also hosts the CSO and the James Clerk Maxwell Telescope (JCMT). The area is known as "Millimeter Valley" for its facilities. A state-of-the-art receiver laboratory was established at Harvard in 1987. The SMA began operation in 2003, with funding from the Smithsonian Institution. The SMA now has eight 6 m diameter antennas; the two additional antennas were the result of a partnership with the Academia Sinica Institute of Astronomy and Astrophysics (ASIAA) of Taiwan. The SMA can make observations at frequencies from 180 to 418 GHz and achieve sub-arcsecond resolution. It continues to operate, a successful Northern Hemisphere counterpart to ALMA. The SMA, CSO, and James Clerk Maxwell Telescope (JCMT) are shown in Figure 2.5, a panorama of "Millimeter Valley" (perhaps more properly "Submillimeter Valley") on Maunakea.

Millimeter Arrays in Japan and Europe

The impetus given millimeter interferometry in the United States by the Barrett Report was matched by two developments abroad, whereby both Japan and Europe built arrays of significant size.

Figure 2.5 Millimeter Valley on Maunakea: far left – the Caltech Submillimeter Observatory (CSO); center – the SMA operations center is the shorter structure to the right of the taller JCMT; and right – the SMA. Courtesy of Jonathon Weintroub, reproduced by permission.

Figure 2.6 The Nobeyama Millimeter Array at the Nobeyama Radio Observatory. Courtesy of Masato Ishiguro, reproduced by permission.

Nobeyama Millimeter Array – In 1982, the National Astronomical Observatory of Japan (NAOJ) completed construction of the Nobeyama Millimeter Array (NMA), which with five 10 m diameter antennas was intended to be the largest millimeter array at that time (Figure 2.6). First observations with the NMA at the Nobeyama Radio Observatory (NRO) were made in 1984. The NMA stopped operation in 2007 when Japan entered the ALMA project. Over the 23 years of its operation the NMA produced over 100 studies of Galactic and extragalactic sources.

Figure 2.7 The IRAM NOEMA Interferometer on the Plateau de Bure, France. Credit: ©IRAM, reproduced by permission.

IRAM Plateau de Bure Interferometer – IRAM installed a millimeter interferometer on a high plateau near Grenoble France in 1982. The interferometer has been extraordinarily productive. The demonstration that infrared-luminous galaxies are merging galaxies is only one of the significant results in many areas that could be cited. The interferometer received a significant upgrade in recent years, the Northern Extended Millimetre Array (NOEMA), shown in Figure 2.7, now consisting of twelve 15 m diameter antennas.

The Barrett Report's advice to NSF and the Foundation's subsequent support of the development of millimeter interferometry in the United States laid the groundwork for the Report's main recommendation, namely, that a large millimeter array was needed to study star formation, galactic structure, and the evolution of galaxies. The concept development of that facility and the proposal to build it, submitted in 1990, is the story of Chapter 3.

Notes

1   Voting for: Bernie Burke (MIT), Dave Hogg (NRAO), and Don Osterbrock (U. California, Santa Cruz); against: Jacques Beckers (NOAO), Riccardo Giacconi (STScI), Fred Gillett (NOAO), Roberta Humpheries (U. Minn.), Dick McCray (U. Colorado), Peter Pesch (SUNY Stony Brook), and Joe Taylor (Princeton U.); abstaining: Eric Becklin (U. Hawaii). From Gordon (2005).

2   No one worked harder to realize the 25 Meter Telescope than Mark Gordon, long-time manager of the very successful 36 Foot, then 12 Meter, Telescope. An account of the

project can be found in his book *Recollections of "Tucson Operations": The Millimeter Wave Observatory of the National Radio Astronomy Observatory* (Gordon, 2005).

3 Nick Scoville (U. Mass.), Paul Vanden Bout (U. Texas, Austin), Jack Welch (U.C., Berkeley), Bobby Ulich (Multi-Mirror Telescope Obs.), Frank Lovas (National Bureau of Standards), Marc Kutner (Rensselaer Polytechnic Institute), Pat Palmer (U. Chicago), Paul Goldsmith (U. Mass.), Jill Knapp (Princeton U.), Ed Churchwell (U. Wisconsin), Alan Barrett (MIT), Pat Thaddeus (Goddard Institute of Space Studies), Tom Phillips (Caltech), Tony Stark (Bell Labs.), and John Bally (Bell Labs.).

4 Wilson to Vanden Bout and others, NAA-NRAO, MMA, MMA Planning, Box 1. https:// science.nrao.edu/about/publications/alma

5 Notes taken throughout the meeting by Solomon, his hand-written draft of the letter to Bautz and Johnson, and a copy of the 29 October 1982 letter can be found at NAA-NRAO, MMA, MMA Planning, Box 1. https://science.nrao.edu/about/publications/alma

6 Bautz to Wilson and Co-Signatories, 23 November 1982, NAA-NRAO, MMA, MMA Planning, Box 1. The members Bautz appointed were Alan Barrett, Chair (MIT); Dennis Downes (IRAM), Charles Lada (U. Arizona); Pat Palmer (U. Chicago); Lew Snyder (U. Illinois, Urbana); and Jack Welch (U.C. Berkeley). Ex-officio: Vernon Pankonin (NSF Staff Liaison) and H. Crismond (NSF), secretary to the committee.

7 *Report of the Subcommittee on Millimeter and Submillimeter Wavelength Astronomy, NSF Astronomy Advisory Committee*, April 1983, NAA-NRAO, MMA, MMA Planning, Box 1. https://library .nrao.edu/public/memos/alma/main/memo009.pdf

8 See David (1983) and Waldrop (1983) for the complete articles.

9 A third interferometer intended for observation of H_2O sources was constructed at MIT by Bernie Burke. Its operation was never realized.

10 Welch (1996)

11 Plambeck (2006)

12 Woody, Vail, and Schall (1994) describe the design, construction, and performance of the Leighton 10 m antennas.

13 See, for example, Scoville, et al. (1986).

14 See, for example, Sargent and Beckwith (1987).

15 Woody, Beasley, and Bolatto (2004) describe CARMA.

3

The Millimeter Array

"Big things have small beginnings."

T. E. Lawrence

First Concept of a US National Array

To understand the origins of the Millimeter Array (MMA), NRAO's precursor project to ALMA, we need to go back to the early development of interferometers working at millimeter wavelengths in the decade preceding the Barrett Report. As was mentioned in the previous chapter, this occurred first at the Hat Creek Radio Observatory (HCRO) of the University of California, Berkeley, simultaneous with the Observatoire de Bordeaux, and a little later at the Owens Valley Radio Observatory (OVRO) of Caltech. The first astronomical observations of this kind were obtained at HCRO. A description of that interferometer, operating at 13.5 mm wavelength, was reported in the Proceedings of the IEEE.[1] At that time, it consisted of two antennas, one of diameter 3 m and another of 6.1 m, spaced 265 m apart. The interferometer was operated using a PDP-8/S minicomputer with data storage of only 4 kbytes of 12-bit memory, which was respectable for the early 1970s. Plans were reported in the article for a larger interferometer capable of operating at considerably shorter wavelengths. That instrument was built with three 6.1 m diameter antennas that could be moved on two perpendicular tracks, providing an angular resolution of 1–2 arcseconds for radiation received at 2 mm wavelength. With this instrument, observations were made of interstellar water vapor sources and planets. High-resolution measurements of the positions of water vapor emission regions constitute the first reported astronomical results from a millimeter-wavelength interferometer,[2] although, strictly speaking, the wavelength of the observations was 1.35 cm.

Figure 3.1 Photograph of the two-element interferometer built at the Observatoire de Bordeaux to observe the Sun. Courtesy of Alain Baudry, © Collection Observatoire Aquitain des Sciences de l'Univers, Laboratoire d'Astrophysique de Bordeaux, reproduced by permission.

Pioneering work on millimeter interferometry was also carried out abroad. An article, published back-to-back with the paper describing the HCRO interferometer in the Proceedings of the IEEE, reported the first operation of a small millimeter interferometer at the Observatoire de Bordeaux,[3] in France. As shown in Figure 3.1, it consisted of two 2.5 m diameter antennas spaced 64.4 m apart and operating at 8 mm wavelength. It was operated using a Honeywell H316 computer having a small memory (8 kbytes) similar to the HCRO instrument. The interferometer was built to make observations of the Sun, and the interferometer's resolution of roughly 20 arcseconds was designed to enable more detailed measurements of active regions in the Solar photosphere, which are typically of 1–2 arcminutes in diameter. Solar observations were first achieved on 29 January 1973. The Bordeaux and Berkeley groups closely cooperated with one another. Jack Welch received an honorary degree from Bordeaux University in 1979.

The international radio astronomy community that coalesced after the Second World War was relatively small, very competitive, but strongly collegial. Consequently, the scientific and engineering staff at the NRAO were aware of these developments well before the publication of the papers in the IEEE Proceedings. Indeed, in the mid-1970s, in a scientific staff meeting called by then director Dave Heeschen, the issue of whether NRAO should build a millimeter wavelength interferometer as a development project was discussed. Heeschen decided it was best for the NRAO to leave this effort to the universities that were already aggressively pursuing millimeter instrumentation. Heeschen was keenly aware that the NRAO had been founded to build and operate radio telescopes that were *beyond* the capability of a single university to fund, build, and operate. He was overheard to say,[4] when declining to compete with individual universities, that *"NRAO should stick to its knitting."*

Nearly a decade later, when Mort Roberts was the observatory director, the discussion in a routine scientific staff meeting was on which major new NRAO telescope could follow the recently completed Very Large Array (VLA). Frazer Owen made the off-the-cuff suggestion of a millimeter wavelength array, by which he meant something more like the VLA, composed of more than just a few antennas, something bigger than the millimeter interferometers being developed at HCRO, OVRO, and Bordeaux. It is the authors' opinion that this suggestion is the very first of all the early concepts that would lead to ALMA. As he thought further about his suggestion, Owen developed the initial concept and wrote it up in a memorandum to the NRAO staff. It eventually became the first in an NRAO series of MMA Memos, where "MMA" stood for "MilliMeter Array," the working name NRAO chose for this new telescope project.[5]

Owen's memorandum presented the outline of an array that might be built in five to ten years. It was not a formal proposal but was intended to stimulate discussion that could lead to a real proposal. By 1982, the HCRO and OVRO millimeter arrays were in the early stages of operation, and at IRAM planning for an array on the Plateau de Bure near Grenoble, France, was well underway. Owen felt that it was time for a US national array operating at millimeter wavelengths. In his original concept, fifteen 10 m antennas with another fifteen 3 m antennas, were distributed in a Y-shaped configuration like the VLA. Antenna separations up to 1 km in length would be supported and the array would operate at frequencies around 115 GHz and, during good weather, up to 230 GHz. Owen proposed construction on the VLA site in order to take advantage of the substantial infrastructure and expertise already available there. His very rough cost estimate was $33 million in 1980 dollars, that is, a project in the price range of the recently abandoned 25 Meter project, not on the $100 million scale of the VLA.

The two frequency bands suggested by Owen were not arbitrary. In the 12 years since its discovery in 1970, interstellar carbon monoxide (CO) had come to be recognized as the most easily observed signpost of interstellar molecular clouds and the localized regions of star formation contained within them. Because interstellar molecular hydrogen (H_2) cannot be observed in cold gas, the demonstration[6] by one of the authors (Dickman) that CO emission is a reliable quantitative proxy for molecular hydrogen (H_2) made radio observations of carbon monoxide the observational leading edge of molecular cloud exploratory studies in the late 1970s and early 1980s. The two lowest energy spectral lines of CO are at 115 and 230 GHz.

On 11–12 October 1982, a workshop was held at NRAO's Green Bank site to discuss future instrumentation for the Observatory. Members of both the scientific and the engineering staffs attended. A number of millimeter wavelength projects were presented, but the discussion centered on millimeter arrays. Siting a single millimeter wave dish in the southern hemisphere was proposed but gained little support. The report of the workshop meeting was subsequently summarized by Director Roberts, in an article in the NRAO Observatory Newsletter.[7] The article reaffirmed the Observatory's goals of completing the upgrade of the 12 Meter Telescope and realizing the VLBA.

The Barrett Committee (as discussed in Chapter 2) was well aware of all these developments at the time of its deliberations. A joint meeting of the Committee with community participants, held at Bell Labs Crawford Hill on 10–11 February 1983, was attended by Owen. He and six others had submitted possible science cases for an MMA to the Committee.[8] Each wrote to his own particular interest: Owen on continuum studies; Charles Lada and John Bally, on the formation of stars and planets; Jack Welch on the molecular component of the Galaxy; an extensive list of projects by Jill Knapp; chemistry studies by Bill Langer; and complex galactic molecules by Lew Snyder. These authors were clearly already thinking of an instrument on a national scale: Lada and Bally assumed an array of thirty 6 m diameter antennas, and Knapp assumed sixteen antennas of the same size. These two straw instruments had only 66 percent and 35 percent, respectively, of the collecting area of Owen's concept, but both comprised significantly more antennas than the millimeter arrays then under development.

A Revised Concept

On 3 March 1983, Roberts sent a memorandum[9] to Owen, who was by then an NRAO scientific staff member, urging him to form a committee, preferably of NRAO scientists, to establish the parameters of an MMA required by

Figure 3.2 First revision of the original MMA concept with the 6 m diameter antennas in a Y-configuration on tracks and the 3 m diameter antennas mounted on a steerable platform. Credit: NRAO/AUI/NSF, CC BY 3.0.

the science to be done. The technical and engineering staff would then translate these requirements into specifications and a price for the array. Owen responded with a memorandum[10] on 16 March 1983 stating that he had already appointed such a committee and urging Roberts to provide significant help from the technical staff. The closing statement of the memorandum was, "*I hope the VLBA does not totally blind the Observatory to the future.*" The tasks Owen had in mind were to define the configuration of the array, that is, the arrangement of antennas on the ground, the signal correlator that compares signals from the antennas, and the site. By "significant help" he was hoping for one or two full-time engineers.

Owen's committee struggled with a conundrum: how to compromise between the number and size of the antennas. Those choices affect the area of the sky seen by the individual antennas, and the sensitivity of the array to widely distributed, comparatively weak sources, such as CO emission from the outer regions of interstellar molecular clouds. To address these issues, Owen's original concept was changed to that shown in Figure 3.2. The 6 m diameter antennas were left in a Y-shaped configuration, but the 3 m diameter antennas were mounted close together on a large steerable platform. Having them close together provided sensitivity to widely distributed sources. It also precluded the antennas from blocking one another's view when observing near the horizon. This revised concept[11] was the largest of the proposals discussed at another NRAO future instrumentation workshop, also held at the Green Bank site, on 20 September 1983.

In spite of all this discussion, the construction of a national MMA – the highest priority project suggested by the Barrett Committee to NSF – was nowhere to be seen in the Observatory's 1983 Long Range Plan. It would be unfairly harsh, however, to assume that the Observatory was indifferent to the Barrett Committee's report. The Report had, after all, been public for only six months. The Observatory was losing no time moving forward with the VLBA – the most mature NRAO project endorsed by the NSF Astronomy Advisory Committee – and what was to become the MMA seems to have been viewed as a more long-range project by most NRAO science and engineering staff. Even so, the Observatory recognized the need for further technical development of the concept, and preliminary design studies of some critical components, like the signal correlator, had begun. A study[12] of the atmospheric stability at 13 mm wavelength had indicated that the VLA site could be used for observations at 3 mm wavelength on antenna separations up to 1 km for 50 percent of the time. The possibility of using the VLA site for the MMA was to remain on the table for another 10 or 15 years, even as the interest in using the array at higher frequencies became more and more incompatible with the atmospheric conditions at the VLA site.

Steps to the MMA Proposal

Over the years, the development of the MMA involved a number of external advisory committees whose reports helped define the concept and laid the basis for an eventual proposal to the NSF for the MMA. News of meetings and results were reported in the NRAO Newsletter and also in the MMA Newsletter, which went into more details on subjects of interest. In addition to the formal reports from the advisory committees and working groups, the MMA Memo Series[13] published memoranda on largely technical subjects by anyone who wished to contribute. The leader of this effort was Owen, who was the driving force in the development of the MMA until Bob Brown assumed leadership of the project in 1985.

Technical Advisory Committee – An external Technical Advisory Committee (TAC) was appointed by NRAO in 1983, chaired by Bob Wilson, with eight more members[14] from various universities as well as Bell Labs. The first meeting of the committee occurred on 1–2 March 1984.[15] The agenda, along with Owen's notes made in preparation for the meeting are contained in MMA Memo #14.[16] The committee's report was published in MMA Memo #16.[17] Among the most significant of the recommendations to emerge from this meeting was advice intended to define better the scientific program of the array. Without knowing how the array would be used, it was difficult for the committee to judge whether the present concept would be well matched to detailed scientific requirements. In fact, it would be another year before the Observatory appointed the external science advisory committee (SAC) that would be needed to develop the detailed case for the MMA's science mission. Nevertheless, some important points were already recognized by the TAC, perhaps first among them being the need to begin working on the design implications of an MMA capable of making large-scale maps of the sky using a technique called "mosaicking." The first generation of millimeter wave radio telescopes had already shown that many galactic sources were so extended that high-resolution interferometric images would have to be stitched together from multiple smaller images. This process is considerably more complicated with interferometric data than making a composite image with an optical telescope, and it was already clear that multiple options for observing protocols and data processing algorithms would need to be researched.

The TAC also advised investigating sites other than the VLA, if only for due diligence. In doing so, the Committee further urged that detailed studies be made of the transparency and stability of the atmosphere above the VLA site, thus providing a quantitative baseline for evaluating other sites. Because it was already developed and linked to existing NRAO infrastructure, the Committee also noted that if the VLA site was found to be acceptable it should be considered as the MMA default location barring changes in the array's scientific mission. With respect to antennas, the committee expressed uneasy feelings based on NRAO's experience with the difficult resurfacing project[18] of the Observatory's 12 Meter Telescope. Carbon fiber technology was thought to be a risky but promising approach to making the high-precision radio antennas that would be required for a millimeter wave array. Superconductor-insulator-superconductor (SIS) receivers were thought to be the most promising technology. Unsurprisingly, given the makeup of the TAC, the report ended with a call for more university involvement in the development of MMA's technology. Sometime after the report was forwarded to the NRAO Director, it was decided to focus on three areas: antenna configurations, the design of the support structure of the central element carrying the small antennas, and atmospheric studies of the VLA site. Interestingly, the suggestion of a site in South America for the MMA was dismissed by the committee as "not an attractive idea."

In late 1984, NRAO staff scientist Mark Gordon wrote a prescient memorandum[19] entitled "Are We Thinking Boldly Enough?" He thought it was absolutely necessary for NRAO to build an MMA that would be unique compared to anything existing or planned elsewhere. The obvious solution to him was to build the array in Chile, which hosted three major optical observatories. Again, this idea was not taken seriously, as those involved at NRAO thought Chile was as too far away from US expertise in interferometry and was likely to add additional construction and operations costs to the project. More than a decade later, the MMA would be an international project requiring operation at frequencies much higher than 230 GHz, and the case for a Chilean location would be considered much more compelling.

Science Advisory Committee – The appointment of the TAC soon made clear the value of involving the US astronomical community in defining the science mission of the MMA. It was also obvious that, along with the TAC, an SAC was needed as a complementary standing body. The SAC's task would be both detailed and broadly contextual since NRAO would eventually need to write a formal proposal to the NSF for funding the MMA. The proposal would have to establish the general scientific discovery space that the instrument would occupy and describe where it would fit into the suite of leading-edge US astronomical instruments, such as the Hubble Space Telescope (HST) and the Keck 10 m (optical) telescopes, already under construction or being planned for the next decade.

The appointment of an external SAC[20] was announced on 1 March 1985. The committee chair was Jack Welch of U.C. Berkeley, one of the leaders in the development of millimeter-wave interferometry. It would have two meetings during the next year that would be open to the astronomical community. The first was to be held in Tucson, Arizona, at the University of Arizona, 9–10 May 1985, following the NRAO Users Committee meeting. The second was scheduled for Charlottesville, Virginia, at the NRAO Headquarters, just before the meeting of the American Astronomical Society at the University of Virginia. The agenda for the latter meeting was published in the NRAO Newsletter #23.[21] At that time, NRAO scientist Al Wootten started the MMA Science Memo Series.

The SAC held a workshop on 30 September–2 October 1985, at the NRAO Green Bank site for the purpose of defining the science goals of the MMA. About 60 attendees were organized into seven working groups chaired by the SAC members. The working groups and chairs were: Solar System (Imke de Pater), Sun and Stars (George Dulk), Evolved Stars and Circumstellar Shells (Phil Schwartz), Star Formation and Molecular Clouds (Neal Evans), Astrochemistry (Lew Snyder), Low-z Extragalactic Studies (Leo Blitz), and High-z Extragalactic Studies (Bruce Partridge). The letter of invitation stated that the reports of the working groups would form "the backbone of the science section of a conceptual proposal to be prepared in 1986." A brief report was published in the November 1985 NRAO Newsletter (#25).[22] The reports of the working groups were announced in the NRAO April 1986 Newsletter (#27)[23] and the reports of each of the working groups were published as the MMA Science Memo Series.[24]

A much more extensive report was given later in *Science with a Millimeter Array – MMA Design Study Volume I*, issued in January 1988.[25] Chapter I presented the MMA concept at that time. There were to be 21 moveable antennas of about 10 m diameter and 21 antennas of about 4 m diameter mounted on a structure about 29 m in diameter. Several configurations were illustrated. Chapters II–VIII gave the reports of the working groups, who generally found their science goals to be satisfied by the concept of Chapter I. These science goals became the basis for the science case of the MMA proposal.

Management Changes Affecting the MMA – On 1 January 1985, one of the authors (Vanden Bout) was appointed the NRAO Director. On 1 May 1985, he announced the appointment of Bob Brown as NRAO Associate Director for Operations, a position that effectively made Brown his deputy. The eventual division of responsibilities had the Tucson and Green Bank sites reporting through Brown to Vanden Bout, along with the Central Development Laboratory in Charlottesville, where NRAO's electronics were developed. The Socorro site and the VLBA Project reported directly to Vanden Bout. This left Brown sufficient time to supervise the MMA development while Vanden Bout worked to secure funding for the VLBA and, eventually the MMA. Later in 1985, it was learned that the State of New Mexico would provide funds to construct an operations center for the VLBA on the campus of the New Mexico Institute of Technology in Socorro. Now called the Pete V. Domenici Science Operations Center, this facility also became the site for VLA operations, which moved from the Plains of San Agustín 60 miles west of Socorro. This development strengthened the vision shared by some NRAO staff members of Socorro becoming the center for all NRAO arrays, including the MMA.

Final Steps toward a Formal MMA Proposal – The first formal report[26] of specifications for the MMA based on identified scientific goals was *Millimeter Array Design Concept – MMA Design Study Volume II*, issued in January 1988 along with Volume I. It refined the concept presented in Volume I in a series of steps. Chapter I restated the recommendations of the Barrett Report. Chapter II went into more detailed requirements from each of the science areas. In Chapter III, the heterogeneous array of 10 m and 3 m diameter antennas had disappeared to be replaced by a homogeneous array. Wide-field images were to be produced by "mosaicking," a technique that stitches together multiple images.[27] Chapter IV discussed practical details, including the selection of a site. Mount Baldy near the VLA site was discussed in particular. It concluded with the parameters of the new design concept: forty 7.5 m diameter antennas of surface accuracy such that deviations from a perfect parabola are less than 0.043 mm on average; moveable into configurations of 90, 300, 1,000, and 3,000 m largest extent; operating in bands of 36–48, 70–115, 200–270, and 270–350 GHz. This concept would provide images with 0.1 arcsecond angular resolution at 230 GHz, equivalent to discerning a basketball hoop at 600 miles distance. In Chapter V, the technique and application of mosaicking were discussed in more detail. It requires the ability to move the antennas quickly, which is a challenge for the antenna design. The principal player here was Tim Cornwell, a scientist at the VLA. He wrote a series of articles on image processing relevant to the MMA, the one on mosaicking appeared as MMA Memo #24. The construction budget was estimated to be $66 million in 1988 dollars including $11 million for contingency. Major construction was to begin in 1992.

MMA Proposal

On 6 October 1986, Frazer Owen distributed draft chapters for an MMA proposal to the NRAO staff for comment. The TAC met later that month to hear updates on site testing and imaging studies and to review the draft proposal. (The alert reader will notice that this is more than a year prior to the publications of the Millimeter Array design volumes which were the basis for the proposal. The contents of those reports were available in draft long before their final publication, which was essentially a formality by 1988.) Refining this material and settling, to the extent possible, as many of the remaining questions took another year. A meeting at NRAO Tucson discussed technical issues on 1 November 1988. A second science workshop was held on 19–23 October 1989 in Socorro, NM. In January 1990, AUI formally submitted the proposal[28] to the NSF. The proposal began with an introduction recalling the requirements stated in the Barrett Report and noting that although millimeter astronomy had been pioneered in the United States, in contrast to other countries, "*no major national millimeter-wave instrument has ever been funded.*"

The proposal called for an array of forty identical 8 m diameter antennas that would provide a total collecting area of 2,010 m^2. Figure 3.3 shows the relative collecting area of the MMA compared with the arrays discussed in Chapter 2 and the area that was later achieved with ALMA. The surface accuracy of the antennas would be sufficient to allow observations at wavelengths as short as 1 mm. That is, the array would be able to observe all three of the lowest frequency transitions of CO: 115, 230, and 345 GHz; these being the anticipated workhorse observational frequencies. The proposed array had four configurations for the antennas. Figure 3.4 illustrates the proposed configurations. One was a compact array whereby all the antennas were placed as close together

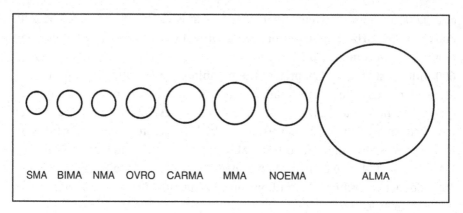

Figure 3.3 The circles indicate the relative geometric collecting areas of the millimeter/submillimeter arrays that have been discussed, for comparison with the proposed MMA, as well as that of ALMA. The actual power of an array depends on additional factors: efficiency of the antenna optics, performance of the electronics, and quality of the atmosphere at the array site. Courtesy of Thomas van den Bout, reproduced by permission.

Figure 3.4 Illustration on the cover of the MMA proposal showing an artist's conception of the array on a high, arid, flat site. Credit: NRAO/AUI/NSF, CC BY 3.0.

as practical. There were three more extended configurations of 250, 900, and 3,000 m in diameter. The angular resolution of the largest configuration was 0.07 times the observing wavelength in arcseconds. At 115 GHz, this is about 0.2 arcseconds, 10 times better than the Barrett Report recommendation. There were to be four observing bands: 30–50, 68–115, 130–183, and 195–366 GHz with the ability to observe in more than one band at a time. In short, the MMA was to be a flexible array, a millimeter version of the VLA.

The proposal's science goals had broadened beyond the observation of interstellar carbon monoxide. These were summarized in the introduction as follows:

1. Image the redshifted dust continuum emission from evolving galaxies at epochs of formation as early as $z = 10$.
2. Reveal the kinematics of optically obscured galactic nuclei and QSOs on spatial scales smaller than 100 pc.
3. Assess the influence that chemical and isotopic gradients in galactic disks have on star formation and spiral structure.
4. Image heavily obscured regions containing protostars, and protostellar and preplanetary disks in nearby molecular clouds, with a spatial resolution of 10 AU and kinematic resolution <1 km/sec.
5. Detect the photospheric emission from hundreds of nearby stars in every part of the Hertzsprung–Russell (H–R) diagram.
6. Reveal the crucial isotopic and chemical gradients within circumstellar shells that reflect the chronology of stellar nuclear processing and envelope convection.
7. Resolve the dust-formation region and probe the structure of the magnetic field in stellar winds.
8. Establish the relative distributions of the large number of complex molecular species in regions of star formation, relating them to shock fronts, grain disruption, and energetic outflows – information that is essential to the understanding of astrochemistry.
9. Obtain unobscured sub-arcsecond images of cometary nuclei, hundreds of asteroids, planetary atmospheres and surfaces, and solar regions of active particle acceleration.

Chapter II of the proposal gave an expanded discussion of the array, particularly of its sensitivity. The fact that the array should detect the continuum radiation from a 0.1 mJy source in one hour of observing time was noteworthy. A graph showed that this placed the MMA in the same class for detection of extragalactic sources as the VLA, the Spitzer InfraRed Telescope Facility (SIRTF) in the infrared, the HST in the optical, and the Advanced X-ray Astronomy Facility (AXAF) in X-rays. A more detailed discussion of the science proposed for the MMA was given in Chapter III: observations of the distant Universe, the Universe nearby, the Sun and stars, molecular clouds and star formation, astrochemistry, evolved stars and circumstellar shells, and planetary science. Chapter IV discussed the techniques for transforming array observations into images. Simulations of these techniques were presented. The proposal did not include a large single dish, thought heretofore to be necessary for accurate imaging. Instead, the chapter concluded with a pledge to continue research into the viability of the mosaicking technique.

Chapter V discussed computing needs, both to operate the array and to produce images from recorded data. The so-called real-time system needed to operate the array was modest in scope and cost, estimated at $200k per year. The systems needed for later image processing were much larger and more expensive. Its components were:

1. A machine with a total computing power of 8 billion operations per second, costing $5 million;
2. a minimum of 300 Gbytes of off-line storage, costing $1 million;
3. about 10 high-performance graphics workstations for a total of $500,000;
4. 40–50 personal workstations for a total of $500,000–1 million;
5. a system of mass storage, costing $1 million;
6. local area networks, costing $500,000; and
7. printers etc., costing $200,000–300,000.

The site selection process and status at the time of the proposal were presented in Chapter VI. The requirement to observe at 1 millimeter wavelength implied a site with less than 2 mm precipitable water vapor overhead for at least some of the time. This meant a high site, at least 2,700 m in elevation, and one large enough to accommodate the 3 km configuration. The search identified 50 sites in the continental United States south of 36 degrees north latitude with this elevation. The latitude requirement was imposed to allow observations of the center of the Milky Way, our own galaxy. Eliminating sites that were too small, that were in national wilderness areas, or had access problems, left only three that were worthy of serious study: two in the Apache National Forest near the towns of Springerville and Alpine, both in Arizona, and one on Mt. Baldy in the Magdalena Mountains near the VLA site.

The proposal illustrated the placement of the 3 km configurations on these sites and presented atmospheric opacity data obtained with a tipping radiometer operating at 225 GHz on the Magdalena Mountain site. The measurements showed that the site had an atmospheric transparency of 85 percent for about half the time. That is, the MMA could operate there although operation at 345

GHz would be limited. NRAO built four of these "tippers." One had operated on Maunakea at the CSO site for several months at the time of the proposal and the data indicated that Maunakea was, indeed, a very dry site, suggesting expanding the site search to Hawaii. The proposal also mentioned studying sites in northern Chile. Sites in both Hawaii and Chile would come at a premium in cost, so the focus at the time was on the US continental sites. The proposal fairly presented the question of where to locate the MMA as an ongoing investigation.

Chapter VII discussed the antennas for the MMA, a subject still under investigation. Making the backup structure that supported the reflecting panels of carbon-fiber-reinforced plastic (CFRP) material was attractive. A structure made of CFRP would be both lighter and stiffer than one of steel. But the technology was relatively new at the time and so the proposal called for a hybrid like that being used at IRAM – a combination of CFRP and steel structural members. Whether the panels would use CFRP technology was left open. The conceptual design of the antenna was a very compact structure, as shown in Figure 3.5.

Despite the uncertainties surrounding the antenna design, a cost estimate was made that came to $1.25 million per antenna. The issue of how the antennas were to be moved from one configuration to another was left for future investigation.

The MMA electronics plan was presented in Chapter VIII. It was clearly stated that the receivers would require development work to meet the specifications. There was a lengthy discussion on the choice of style for the signal correlator – one like the VLBA or one like the VLA. The VLA type was chosen as being less expensive, of order $6.2 million. Chapter IX estimated operations costs on the assumption that the MMA would be located at one of the continental US sites, that is, not too far from the VLA. With a staff of 90, the annual cost to operate the MMA was estimated to be $6.6 million. Chapter X presented a summary of all the estimated component costs that totaled $104.3 million. A contingency of $15.7 million (15 percent) was added to make a grand total of $120 million. It must be noted that at the time NRAO was not running construction projects with the management tools that are now standard for large projects. This total cost estimate was not built bottom-up with a budgeted contingency based on the sum of contingencies for each component. In fact, there was an overriding consideration that went into the cost estimate – that the total request does not exceed the inflated cost of other NRAO projects like the VLA and VLBA. It all added up to a number that was judged to be one the NSF could swallow. The schedule presented called for three years of development, starting in 1991 at $1 million and continuing in 1992 at $2 million and in 1993 at $5 million.

On 26 July 1990, Bob Hughes, President of Associated Universities, Inc. (AUI), the research management corporation that operates NRAO for the NSF,

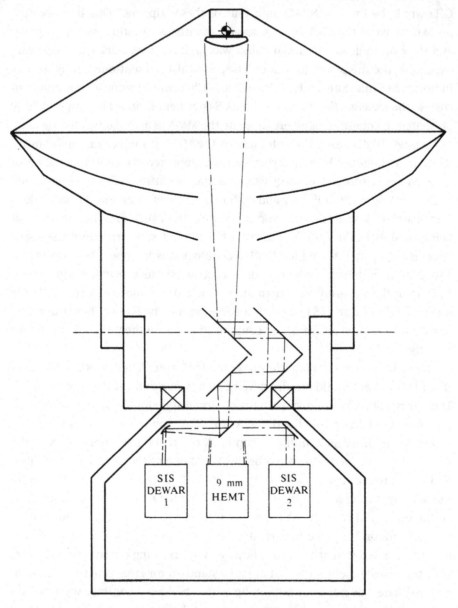

Figure 3.5 Conceptual MMA antenna design. Plane mirrors guide the beam to one of three cryogenic dewars, two with SIS mixer receivers and one with a high electron mobility transistor (HEMT) amplifier receiver. Credit: NRAO/AUI/ NSF, CC BY 3.0.

formally submitted the proposal to the NSF, where it was assigned the proposal number AST-9024403. The request caught the attention of the science press. An article[29] by Robert Crease in the 28 September 1990 issue of *Science* began with the following paragraph. "*In an era of Gramm-Rudman budget cuts, shrinking research grants, and post-Hubble sensitivities, it takes more than a little chutzpah to propose yet another big science project, especially one in a relatively new field. But that is what a stalwart band of astronomers, led by Robert Brown of the NRAO, has just done."* The article went on to trace the history of US millimeter astronomy, note the foreign competition, and end on a somewhat positive note: *"The 25th anniversary of the discovery of the 2.6 mm carbon monoxide line in space will be in 1995. By then astronomers hope to be on the verge of seeing the first glimmers of mm light through the Millimeter Array."*

Proposal Review

In her letter[30] acknowledging receipt of the proposal, Pat Bautz, AST Director, informed Hughes that Vernon Pankonin would be in charge of its review. On 2 August 1990, in response to a request from the NRAO Program Officer, Ludwig Oster, Brown sent 37 names of suggested referees, of which 16 were highlighted as being "particularly knowledgeable." The 25 anonymous referees who eventually submitted reviews must have included some of those on Brown's list.

While the referees were considering the proposal, Pankonin convened a 10-member committee to discuss the proposal, with a so-called "site visit," held at NRAO in Socorro, NM, on 9–10 April 1991. The committee was chaired by Charles Townes of the University of California (Berkeley). A number of NSF officials attended the proceedings. The meeting agenda consisted of short presentations on the array concept, project organization, imaging, site options, antenna design, and electronics. Most committee discussions focused on the number of antennas and sensitivity required to accomplish the science goals. Another concern of the committee was the fraction of the time at the sites being considered for which the phase stability of the atmosphere was adequate for observations with the proposed configurations. Owen asserted that for Mt. Baldy the phase stability at 225 GHz was 15 degrees or better for 30 percent of the time, one radian or better 60 percent of the time.[31] If the committee wrote a report, NRAO never received it. Instead, a letter[32] to NRAO from Pankonin on 4 November 1991 summarized the concerns of the referees.

Pankonin's letter gave the selection of a site as the most common concern of the referees. They noted that the sites under consideration were all within a half-day drive of the VLA and wondered if this was merely coincidence. They recommended more studies of the three sites with particular attention to phase stability. The referees were split on whether a 15–20 m diameter single dish was required for good imaging. The majority favored the proposed uniform array of forty 8 m diameter dishes. There was concern over the lack of any management structure. The Joint Development Group (JDG), whereby MMA development work could be funded at university groups, was considered an important step. There was concern that the costs were uncertain, particularly, those for computing and operations. These concerns aside, there was general enthusiasm expressed by the referees for the MMA, and Pankonin urged attention to the issues they raised. He asked for an initial NRAO response before he concluded his own analysis. The response[33] was sent on 13 November 1991.

On 10 March 1992, Julie Lutz, who succeeded Bautz as AST Director, sent a letter to AUI President Hughes stating that the review process had been concluded, and giving the NSF's status for the project. She noted that the MMA had been endorsed by the National Academy of Sciences Survey of Astronomy for the 1990s, the Bahcall Report, as one of two top priorities for ground-based astronomy.[34] Her own advisory committee, the ACAST, had endorsed it as well. She said that the NSF considered the proposal to be a detailed but preliminary description of the project and requested that NRAO submit a detailed plan for research and development. She hoped to begin a research and development phase in the fiscal year (FY) 1994. A decision on actual construction would occur midway in the research and development phase, for which NRAO would need to submit a detailed construction plan. The letter enclosed the 25 (anonymous) referee reports[35] and the ACAST resolution.

Design and Development Plan

NRAO's MMA Design and Development Plan[36] was submitted to NSF on 31 August 1992. The plan presented a four-year program, 1993–1996, to complete the tasks that would lead to the start of construction in 1997. The approach was to evaluate technical alternatives, construct and test prototypes, build a single-baseline set of electronics, design and build a prototype antenna, and prototype software for evaluation by future users. This work was to be conducted in concert with the university groups in the JDG, represented at the time by Jack Welch (U.C. Berkeley), John Carlstrom (Caltech), Lew Snyder (U. Illinois), Leo Blitz (U. Maryland), and Paul Goldsmith (U. Mass.).

Lest anyone think the MMA concept had gone stale in the two years since the submission of the proposal, the tone of the document's introduction was that of a sales pitch. It ends by noting that all new generation telescopes, such as the Keck Telescopes and the Green Bank Telescope, required advances in technology, an allusion to the need for design and development funding if the MMA was to succeed. The cover of the plan, shown in Figure 3.6, was impressively high-tech for the time. It illustrated progress in electronics that had already been made: a multi-beam receiver that had recently been put in use on NRAO's 12 Meter Telescope.

Figure 3.6 The illustration on the cover of the first MMA Design and Development Plan was of a multi-beam receiver for the NRAO 12 Meter Telescope. The six receiver cartridges were cooled by a common refrigerator and held in a common cryogenic container, a scheme later used for ALMA. Credit: NRAO/AUI/NSF, CC BY 3.0.

In addition to the ongoing program of atmospheric monitoring being conducted on the three continental US sites, the plan mentioned monitoring at the CSO site on Maunakea. But the latter was not proposed as a possible site for the MMA. Site tasks included biological assessments; the impacts on operations from staffing, power, and communications, roads and snow removal, difficulty in moving antennas; comparisons of imaging with array configurations possible at the three sites; and community considerations. Environmental studies would be done before giving a priority ranking to the three possible sites.

Reasonably detailed descriptions of the development tasks for the antenna and electronics were presented. The staffing requirements for the entire Design and Development phase rose from 10.5 work years in 1993 to 41.4 in 1994, 51.8 in 1995, and 50.3 in 1996. The total cost of the plan was $22.3 million,

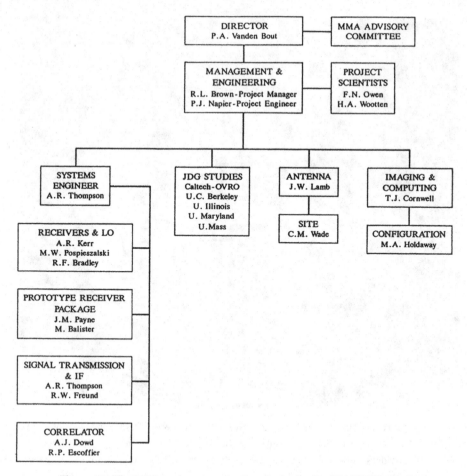

Figure 3.7 The MMA project organization chart. Credit: NRAO/AUI/NSF, CC BY 3.0.

including $2 million in new test equipment, and a simple work breakdown structure showed how the budget should be spread over four years. Criticism that the project lacked a management structure was addressed with an organization chart shown in Figure 3.7. The management structure for the program was that submitted earlier to NSF. At the end of the plan, progress already achieved was documented in Appendices, 15 in total, that were reprints from the MMA Memo Series.

It would be two years before the National Science Board (NSB) approved the MMA Design and Development Plan. In the intervening years, Hugh van Horn became AST Director and Bob Dickman was appointed Head of the Radio Astronomy Unit. Two years later, Dickman presented the Plan to the NSB and gained its approval in November of 1994. Funding of the Plan had to wait until May of 1998 when it could be accommodated by the Major Research Equipment

(MRE) account. NRAO spent the six years from submission of the Plan until its funding refining the science goals and engineering requirements and, most significantly, in the selection of a site. Of course, the proposed start of construction in 1997 had already passed by this time.

Notes

1 Hills et al. (1973) give a technical description of the HCRO two-element interferometer.

2 Hills et al. (1972) report the positions of galactic water emission sources.

3 Delannoy, Lacroix, and Blum (1973) give a description of the U. Bordeaux solar interferometer.

4 T. Riffe and D. Hogg to Vanden Bout, private communication. Both Riffe and Hogg recall Heeschen using this phrase whenever he rejected initiatives that he thought conflicted with or distracted from NRAO's mission.

5 Owen, F. 1982, *The Concept of a Millimeter Array*, MMA Memo #1, http://library.nrao .edu/public/memos/alma/main/memo001.pdf.

6 Dickman (1978) established the correlation between the CO emission from a molecular cloud and the optical opacity caused by interstellar dust, and thence, to the mass of molecular hydrogen in the cloud.

7 Roberts, M. 1982, Future Instrumentation in Radio Astronomy, NRAO Newsletter #9, http://library.nrao.edu/public/pubs/news/NRAO_NEWS_9.pdf.

8 Owen, F. 1983a, *Science with a Millimeter Array*, MMA Memo #2, http://library.nrao.edu/ public/memos/alma/main/memo002.pdf.

9 Roberts to Owen, 3 March 1983, NAA-NRAO, MMA, MMA Planning, Box 1.

10 Owen to Roberts, 16 March 1983, NAA-NRAO, MMA, MMA Planning, Box 1.

11 Owen, F. 1983b, *Concept of a Compound Millimeter Array*, MMA Memo #10, http://library .nrao.edu/public/memos/alma/main/memo010.pdf.

12 Sramek, R. 1983, *VLA Phase Stability at 22 GHz on Baselines of 100 m to 3 km*, VLA Test Memo #143, http://library.nrao.edu/public/memos/alma/main/memo008.pdf.

13 All the memoranda for the MMA and ALMA are available on the NRAO Library website. https://library.nrao.edu/allalma.shtml.

14 The membership of the TAC included Paul Goldsmith (U. Massachusetts), Al Moffet (Caltech), Pat Palmer (U. Chicago), Tom Phillips (Caltech), Larry Rudnick (U. Minnesota), Tony Stark (Bell Telephone Labs.), Bobby Ulich (U. Arizona), Jack Welch (U.C. Berkeley), and Bob Wilson (Bell Labs.) chair.

15 Owen, F. 1983c, Millimeter Array Status, NRAO Newsletter #16, http://library.nrao.edu/ public/pubs/news/NRAO_NEWS_16.pdf.

16 Owen, F. 1984, *Notes on Presentations at the Meeting*, MMA Memo #14, http://library .nrao.edu/public/memos/alma/main/memo014.pdf.

17 Wilson, R. 1984, Report of the Millimeter Array Technical Advisory Committee on their conclusions as a result of the meeting on 1 and 2 March 1984, MMA Memo #16, http://library.nrao.edu/public/memos/alma/main/memo016.pdf.

18 See Gordon (2005) and Kellermann, Bouton, and Brandt (2020), p. 550, for discussions of the re-surfacing of the 12 Meter Telescope.

19 Gordon, M. 1984, *Are We Thinking Boldly Enough?* MMA Memo #25, http://library.nrao.edu/public/memos/alma/main/memo025.pdf.

20 The announcement of the appointment of the SAC was made by Frazer Owen, Upcoming Open Discussions of the Millimeter Array Project, NRAO Newsletter #22, http://library.nrao.edu/public/pubs/news/NRAO_NEWS_22.pdf. The membership of the SAC included Leo Blitz (U. Maryland), Neal Evans (U. Texas, Austin), George Dulk (U. Colorado,) Fred Lo (Caltech), Bruce Partridge (Haveford College), Imke de Pater (U.C. Berkeley), Lew Snyder (U. Illinois, Urbana), Jack Welch (U.C. Berkeley), chair; NAA-NRAO, MMA, MMA Planning, MMA Advisory Committees.

21 Owen, F. 1985b, Millimeter Array Discussion, NRAO Newsletter #23, http://library.nrao.edu/public/pubs/news/NRAO_NEWS_23.pdf.

22 Wootten, A., 1985, Millimeter Array Science Workshop, NRAO Newsletter #25, https://library.nrao.edu/public/pubs/news/NRAO_NEWS_25.pdf.

23 Wooten, A., 1986, Millimeter Array Science Workshop, NRAO Newsletter #27, https://library.nrao.edu/public/pubs/news/NRAO_NEWS_27.pdf.

24 The science working group reports can be found at: https://library.nrao.edu/mmas.shtml.

25 Wootten and Schwab, eds. (1988). NAA-NRAO, MMA, MMA Planning, Box 2.

26 Brown and Schwab, eds. (1988). NAA-NRAO, MMA, MMA Planning, Box 2.

27 Cornwell gives an early description of mosaicking in MMA Memo #32: https://library.nrao.edu/public/memos/alma/main/memo032.pdf.

28 Brown, R. (1990). NAA-NRAO, MMA, MMA Planning, Box 2.

29 Crease (1990).

30 Bautz to Hughes (undated, ca. July/August 1990), NAA-NRAO, Director's Office, Director's Office Correspondence, NSF Correspondence.

31 The time between reception of a signal by one antenna and that of another is the phase difference. Interferometer images require precise information of these phase differences for image construction. The atmosphere also delays signals. The same delay over all antennas is of no consequence to image formation, but differing atmospheric delays impair it. Phase is measured in degrees and/or radians. The numbers Owen quoted are indicative of a marginally acceptable site.

32 Pankonin to Vanden Bout, 4 November 1991, NAA-NRAO, MMA, MMA Planning, Box 3. https://science.nrao.edu/about/publications/alma.

33 Vanden Bout to Pankonin, 13 November 1991, NAA-NRAO, MMA, MMA Planning, Box 3. https://science.nrao.edu/about/publications/alma.

34 By happenstance, no other large radio astronomy projects had been proposed to the Survey. The Radio Astronomy Panel of the Survey, chaired by K. Kellermann of NRAO, was easily able to identify the MMA as their top choice for a large radio project.

35 Anonymous Reviews of the Millimeter Array Proposal [1991], NAA-NRAO, MMA, MMA Planning, Box 3. https://science.nrao.edu/about/publications/alma.

36 Millimeter Array Design and Development Plan, September 1992, NAA-NRAO, MMA, MMA Planning, Box 3. https://science.nrao.edu/about/publications/alma.

4

Searching for a Site

"In Xanadu did Kubla Khan, a stately pleasure-dome decree."

Samuel Taylor Coleridge

Millimeter Array Site Requirements

The power of a radio telescope depends on its size, the quality of its electronics, and its location. The site requirements depend on the wavelength at which the telescope is to operate and on the nature of the telescope – whether it is a single dish or an interferometer. One of the most important considerations for the MMA site was the transparency of the atmosphere, the extent to which radio waves pass through without being substantially absorbed or diverted. While at the longest wavelengths applicable to radio astronomy this is less stringent, for the millimeter band it is critical. Longward of 1.3 cm wavelength, the atmosphere is clear until one reaches a wavelength of about 1 m, where the ionosphere reflects radio waves. (Amateur radio operators communicate around the globe by reflecting their signals off the ionosphere.) At 1.3 cm, corresponding to about 23 GHz in frequency, water vapor in the atmosphere absorbs radio waves, rendering the atmosphere slightly opaque. At increasingly higher frequencies and correspondingly shorter wavelengths, in the millimeter band, other atmospheric constituents come into play: oxygen, ozone, and again water. Between the broad, opaque spectral lines of these atmospheric constituents, the bands of transparency, shown in Figure 4.1, are the "windows" through the atmosphere in which millimeter/submillimeter astronomy can be conducted. The receiver bands implemented in ALMA were designed to match these windows.

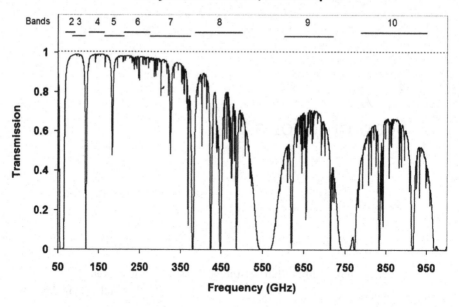

Figure 4.1 The transmission of radiation through the atmosphere as a function of observing frequency (GHz) for 0.25 mm of precipitable water vapor (PWV). The plot is from a model atmosphere for the Chajnantor ALMA site at an elevation of 5,000 m. The ALMA observing bands are indicated at the top for Bands 2–10. (Band 1 covers frequencies around 30 GHz, not shown in this plot.) A PWV column of 0.25 mm is typical of the best observing conditions on the ALMA site. Courtesy of Juan Ramón Pardo, reproduced by permission.

The windows become more transparent as site elevation increases and the concentrations of oxygen and water vapor decrease. Another consideration is that the atmosphere at the site should be stable, that is, should not vary in transparency from one interferometer antenna to another. The flow of air over the interferometer should be smooth, not turbulent. Windy sites are less desirable in this respect. Finally, the site should be sufficiently remote to minimize radio frequency interference. Radio observatories often use the catchy descriptor, "high, dry, and quiet" to describe the ideal site. In addition to a remote, high-elevation site, the MMA required an area sufficiently large and flat to accommodate the array and allow for repositioning the antennas. Ideally, there should be connections to electrical power and water, good access by road, a supportive local community, an available workforce, and a stable government. It took astronomers many years to find the location for their pleasure-dome at what is now the ALMA site in Chile.

United States Continental Sites

In his 1982 MMA concept proposal,[1] Frazer Owen characterized the VLA site as equal or superior to any of the existing millimeter observatory sites at that time, with the exceptions of Maunakea in Hawaii and White Mountain in California. Locating the MMA at the VLA site would also allow the project to benefit from the extensive infrastructure that had been built up there as well as from the considerable expertise in radio interferometry among the VLA staff. The choice of operating frequencies was also a factor and in the early 1980s, the array concept called for receivers operating at 70–120 GHz and over a band centered at 230 GHz, to allow for observations of emission in the two lowest frequency CO transitions. Observing the spectral lines of carbon monoxide was a prime interest of millimeter astronomers, as was discussed in Chapter 1. While this provided a first basis for site selection, over time the desired frequency range became more ambitious.

In June 1987, Mark McKinnon published a report[2] that presented the results of atmospheric opacity measurements at a frequency of 225 GHz with two different instruments on both the VLA site and the top of Mount Baldy in the nearby Magdalena Mountains. The measurements were made in late 1986 to early 1987. At an elevation of 3,500 m, Mt. Baldy is significantly higher than the 2,120 m elevation of the VLA site. With opacities close to 0.03, Mt. Baldy was superior to the VLA site where the lowest opacity was 0.1. These opacities correspond to transmissions through the atmosphere of 97 percent and 90 percent, respectively. However, since these values were seen only in the best months of winter indicating that observations of the CO line at 230 GHz at the VLA would be restricted to the coldest winter nights, with little or no wind; on Mt. Baldy, the times adequate for such observations were more frequent.

Although Mt. Baldy was the first MMA site to be seriously considered, there was motivation for the search to cast a much wider net, first in the continental United States, then in Hawaii, and finally in Chile. Cam Wade, a member of the NRAO scientific staff at the VLA, had evaluated prospective telescope sites for the VLA and was doing the same for the VLBA network of antenna stations. He took on the task for the MMA as well. Wade[3] began by studying topological maps for the southwestern United States looking for areas with an elevation above 2,700 m and south of 36 degrees N latitude. The latitude limitation was to ensure visibility by the MMA of the Galactic Center, and to avoid adverse winter weather conditions associated with more northern latitudes. Suitable candidate sites had to be relatively flat over an area of 3 km north–south by 2 km east–west to accommodate the planned configurations of the MMA. These principal criteria and considerations of accessibility and availability led to only two possibilities in the United States: the Magdalena Mountains (including Mt. Baldy) of New Mexico and the White Mountains of Arizona. The White Mountains have large flat areas above 2,700 m elevation, with good access by paved road from the towns of Springerville and Alpine. Five other initial candidate sites were too close to human activity with its accompanying radio frequency interference, or too remote, or had excessive snowfall: South Park, Colorado, near Colorado Springs; the Cannibal Plateau in the Gunnison National Forest, Colorado; the Grand Mesa, near Grand Junction, Colorado; the Aquarius Plateau, in the Dixie National Forest, Utah; and the Osier Plateau, near Chama, New Mexico.

South Baldy, the more western of Mt. Baldy's two peaks, is the site of the Langmuir Research Laboratory, a research arm of the New Mexico Institute of Mining and Technology that studies lightning. A dirt road leading to the Langmuir Lab provides access to the summit in summer, the season for thunderstorms. For year-round access to the MMA, the road would need to be improved and plowed to keep it free from snow. But that was a minor problem compared to those presented by the topography of the peak. Unlike the flat plain on which the VLA sits, the peak area has steep variations in elevation. Flat terrain existed that could accommodate the smaller configurations of the MMA, but moving the antennas to greater separations, especially to the north, would be daunting. A report[4] by an engineering consultant discussed the options and gave very roughly estimated costs. The problems were summarized in a memorandum[5] by Peter Napier to Brown on 16 May 1991, where he stated, "*If an array larger than 2 km in the N-S dimension is an absolute requirement for the MMA we should not commit to the Baldy site until we have done significant additional engineering and cost studies.*" For this reason and to make sure a better site had not been overlooked, studies of two sites in the White Mountains of Arizona, "Springerville" and "Alpine," were begun. The Alpine site was located just north of the town of Alpine on the south side of Escudilla Mountain. The Springerville site had a larger extended flat area just to the north of Big Lake, about a 30-minute drive from the towns of Springerville and Eager.

In 1990, when the MMA proposal was written and submitted to the NSF, studies of the Springerville and Alpine sites were in the beginning stages. Array configurations for all three sites were illustrated in the proposal, and are shown here in Figure 4.2. The Springerville site allowed for a configuration similar to the ideal one in the artist's impression of the MMA that appeared on the proposal cover (shown in Figure 3.3). The Alpine site came close in this regard, while the Mt. Baldy site looked nothing like the ideal configuration. Even so, imaging simulations showed the Mt. Baldy configuration provided roughly comparable quality, provided the extensive site work required for access to the more distant antenna locations could be made. Given all this, Mt. Baldy was the

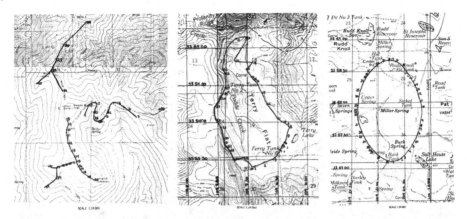

Figure 4.2 The MMA array configurations on the Mt. Baldy (left), Alpine (center), and Springerville (right) sites. The antenna locations are shown as black dots along tracks over which the antennas would be moved. The areas are approximately 3 km east–west by 5 km north–south. Credit: Adapted from topographic maps of the US Geological Survey; NRAO/AUI/NSF, CC BY 3.0.

leading contender for an MMA site in 1990 because atmospheric quality studies had already shown it to be an acceptable site, whereas such studies had not yet begun for the White Mountains sites.

The original MMA proposal did not specify a site. Rather, it only stated that the plan was to continue studying the three continental US sites, eventually pick one of them, and then get permission to begin construction. All of this was to be accomplished by 1994. Contacts with the National Forest Service and local government officials during the investigation of the Alpine and Springerville sites led to publicity in the Arizona media and interest in the nearby communities. A joint resolution[6] was passed in the Arizona Legislature endorsing the construction of the MMA in Apache County. Environmentalists worried about the impact the MMA would have on the Alpine site as it was a habitat for the Spotted Owl. Escudilla Mountain near the Alpine site had been the home of the last grizzly bear seen in Arizona. NRAO was urged to favor the Springerville site, if they chose Arizona.

Several environmental and geographic factors were at play now, and an overriding consideration continued to be the atmospheric transparency. Atmospheric testing was accomplished using radiometers that measured the sky transparency at various elevations above the horizon, tipping the line of sight from the zenith to near the horizon. The radiometers came to be called "tippers." Figure 4.3 shows an open tipper. They operated at a frequency of 225 GHz, near the frequency of the second lowest spectral line of CO. Eventually, four such tippers were constructed by the NRAO Central Development

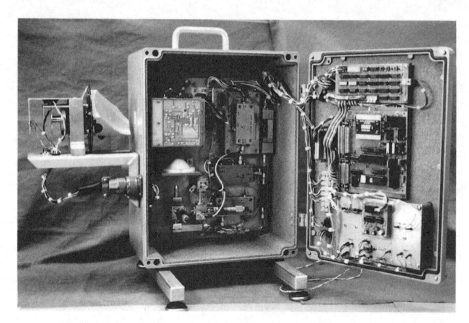

Figure 4.3 An open view of an MMA tipper showing the receiver electronics and computer control and communications circuits. The moving parabolic surface mounted on the side rotates (tips) in elevation to make the observations. The radiometer front end, or receiver, is in the left half of the box, and the intermediate frequency and digital sections are in the right half. The box can be closed for ease of shipment and to preserve temperature stability during operation. Credit: NRAO/AUI/NSF, CC BY 3.0.

Laboratory (CDL). Sandy Weinreb, the lead engineer of the CDL, designed and supervised the tipper construction,[7] along with Z.-Y. Liu, a visiting engineer, and Scott Foster, an NRAO summer student who greatly improved the software.

The first three tippers were successively placed at the NRAO 12 Meter Telescope on Kitt Peak near Tucson AZ, on Mt. Baldy, and at the Springerville site, shown in Figure 4.4. The latter tipper was later moved to the CSO on Maunakea in Hawaii. The tippers measured both atmospheric opacity at 225 GHz and atmospheric stability which could be inferred from the fluctuations in opacity versus time. The MMA proposal noted that the atmospheric stability on both Maunakea and Mt. Baldy were very good. An analysis of the tipper data for Mt. Baldy showed that for 25 percent of the time the opacity was less than 0.081 (92 percent transparency), meeting the specification for a good millimeter observing site of opacity less than 0.1 (90 percent transparency) for a reasonable amount of the time. An analysis of the duration of low opacity periods of time showed that the average was 24 hours, four times the expected typical observation of six hours.

Figure 4.4 The weather station on the Springerville site. The shed held the tipping radiometer and recording equipment. Battery power was maintained by solar panels and a windmill. Credit: Frazer Owen, NRAO/AUI/NSF, CC BY 3.0.

The tipper data confirmed what was indicated by theoretical models of the atmosphere – the elevation of a site is the prime characteristic controlling the transparency. South Baldy at 3,240 m was better than the VLA at 2,120 m, the Springerville site at 2,800 m, and the 12 Meter Telescope site on Kitt Peak at 1,895 m. On Maunakea, the CSO site at 4,072 m was better than the VLBA site at 3,725 m, and both were better than South Baldy. As a result, the MMA Project had to give serious consideration to a non-continental site, namely Maunakea on the Big Island of Hawaii.

Hawaii

Maunakea, on the Big Island of Hawaii, is arguably the best astronomical observing site in the northern hemisphere. At the time of the MMA project's interest in Maunakea, it hosted a number of optical and infrared telescopes, the CSO, and the westernmost antenna of the VLBA. Use of the mountain top for astronomy was managed by the Institute for Astronomy (IfA) of the University of Hawaii, Manoa, under a lease agreement with the State of Hawaii. NRAO had been through the required negotiations, agreements, and permits for a VLBA antenna that had been placed on Maunakea. A sticking point for the MMA was the requirement the IfA imposed on all telescopes sited on Maunakea, namely that the IfA receive 15 percent of the observing time.[8] NRAO was opposed to the granting of a block of time to another organization that would control its use but, if necessary, might consider doing so for no more than 10 percent.

A negotiation with the IfA for an MMA site never occurred. It was not easy to fit the MMA onto the mountain. Don Hall, the IfA Director, thought it could be done and had indicated in broad brush how he would do it. Cam Wade thought it might barely fit onto a flat area stretching east from the VLBA antenna. Nevertheless, NRAO continued to discuss the possibilities with the IfA if only to keep open the possibility that the MMA might be merged with the Japanese Large Millimeter/Submillimeter Array (LMSA), a project[9] with similar science goals to the MMA. The Japanese astronomers already had a large optical/infrared telescope on Maunakea, the Subaru Telescope, and siting the LMSA there as well would offer economies of operation.

Atacama Desert of Chile

The Atacama Desert in northern Chile arguably has the best observational sites for ground-based astronomy in the world. It is to the southern hemisphere what Maunakea is to the northern. But in the Atacama, there are even fewer clouds. As shown in Figure 4.5, moisture from the Pacific Ocean is inhibited from coming onshore by the cold Humboldt Current, which condenses the moisture off-shore. The Andes mountain chain stops moisture from the Amazon Basin from flowing west. The result is a desert where rainfall is rare[10] and the landscape is arid. As such, it has attracted numerous astronomical observatories; among the most prominent optical telescopes at the time of the MMA site search were NSF's Cerro Tololo Inter-American Observatory (CTIO), the Carnegie Institution's Las Campanas Observatory (LCO), and ESO's La Silla and Paranal Observatories. It is also where the MMA, to become ALMA, would eventually be sited.

As was mentioned in Chapter 2, Mark Gordon was the first[11] to suggest locating the MMA in Chile, in his memorandum of 1 October 1984. The reasons for ignoring or dismissing this suggestion were several. Despite the fact that NOAO had, with NSF funding, successfully operated CTIO for many years, *radio* astronomers thought of Chile as being very far away, too far from the expertise required for the MMA. The first MMA Advisory Committee had even dismissed the idea as "*uninteresting.*" In time there were exceptions: Pat Thaddeus installed a "Mini" telescope at Cerro Tololo to map CO in the southern hemisphere.

Figure 4.5 A picture looking south at the Atacama Desert in northern Chile, one of the most arid places on Earth, taken from the International Space Station by Swiss astronaut Claude Nicollier, showing the location of the site selected for the MMA, and in time ALMA. The moisture trapped in off-shore clouds by the cold Humboldt Current are clearly seen from this perspective, as are the clouds in the Amazon Basin blocked by the Andes. The swath of gray that crosses the image diagonally is the Atacama Desert. Credit: Claude Nicollier; ESO, CC BY 4.0.

And, under the leadership of Roy Booth, the Onsala Space Observatory in Sweden built and successfully operated a submillimeter telescope, the Sweden-ESO Submillimetre Telescope (SEST) at ESO's La Silla Observatory. Thaddeus and Booth reported the observing conditions for millimeter astronomy to be superb. By February of 1994, it had become clear that Chile deserved a look. Importantly, Dickman at NSF was strongly encouraging consideration of a Chilean site for the MMA. Vanden Bout and Brown decided to make an exploratory trip, scheduled for May 1994.

Figure 4.6 Left panel: Roy Booth and Masato Ishiguro on a visit to the Vega Valley near Paranal in February 1992. Courtesy of M. Ishiguro, used by permission. Right panel: The LMSA site search team enjoying a meal. They appear to be camping out. (left to right) Nick Whyborn, Wolfgang Wild, Angel Otárola, Kimiaki Kawara, and Naomasa Nakai. Courtesy of Nagayoshi Ohashi, reproduced by permission.

NRAO was by no means the first to search for a site in Chile for a millimeter/submillimeter array. As early as 1991, Roy Booth and Lars Bååth searched the Atacama Desert to identify possible sites for a millimeter array.[12] Before selecting Maunakea for the SMA, the SAO had searched for possible sites in Chile. Using topographic maps, Phillippe Raffin and Alan Kusonuki picked out 22 high elevation places which they visited. They identified a site near the town of Ollagüe, at 4,650 m elevation as the most promising. Their list did not include the high plateau above San Pedro de Atacama that was to become the ALMA site. A team[13] from ESO and Japan, shown in Figure 4.6, visited 20 potential sites[14] for the LMSA in 1992, including some of the sites identified by Raffin and Kusonuki in their 14 May 1992 report.[15] In May 1993, three[16] from the team, led by Angel Otárola, went to the high country east of San Pedro where they were able to view from the side of Cerro Toco the plateau below that was to become the ALMA site. By bad luck they were prevented from descending to the plateau due to snow and penitentes,[17] the icy snow formations found at high altitudes.

The visits by Brown and Vanden Bout to observatories in Chile went smoothly, facilitated by the gracious hospitality of their hosts and the considerable help they gave in making travel arrangements. Peter Shaver, an ESO scientist, was observing at La Silla and gave them a tour of the telescopes, introducing them to the La Silla Director, Daniel Hofstadt. The New

Technology Telescope was particularly impressive, as was the cafeteria, a first-class eatery managed by Herr Shumann, the most famous executive chef in astronomy. Miguel Roth welcomed them to the nearby LCO, which looked to Vanden Bout to be run much like McDonald Observatory in Texas. The chief conclusion from the visits was that these three major observatories had operated successfully in Chile for many years with little trouble, despite political upheavals. Malcolm Smith, the CTIO Director, made time available for a discussion of what it is like to operate in Chile, and he offered a tour of the telescopes even though the groundbreaking ceremony for Gemini South was about to happen. Vanden Bout recalls asking the contractors removing the top of Cerro Pachón for the new telescope what permits were required for such work. A man put a hand on a bulldozer and said, "*This is my permit.*" (The statement gave a totally false impression of the environmental consciousness in Chile; ALMA would undergo a rigorous environmental review and permitting process.) Dick Malow, of AURA, the research management corporation that operates the NOAO, had arrived early for the ceremony and took Brown and Vanden Bout to dinner in La Serena. He was to follow the ensuing development of the MMA and ALMA in Chile with keen interest, making sure no missteps were made that might adversely impact AURA's interests.

The NRAO team was expanded for the trip to the high sites. Hernán Quintana had been an NRAO postdoctoral fellow after graduating from Cambridge University, and was now on the faculty of the Pontificia Universidad Católica de Chile (U. Católica). Brown welcomed Quintana, as a former colleague to the team. Riccardo Giovanelli of Cornell University was invited as a trusted friend and US colleague who was fluent in Spanish and understood South American culture, having grown up in Argentina and Bolivia; Vanden Bout spoke no Spanish, and Brown's high school Spanish was very rusty. Finally, and most important, Leo Bronfman, an astronomer at the Universidad de Chile (U. Chile), asked for the help of Angel Otárola, an expert on telescope site requirements in general. Roy Booth, the director of the SEST where Otárola was working, graciously agreed to the request. Otárola knew the high Atacama Desert exceedingly well, having explored it at length. He also knew people in the many villages and larger towns throughout the desert. His knowledge would prove to be crucial to the MMA site selection.

The expedition began in the coastal city of Antofagasta, with a tour of ESO's Paranal Observatory about two hours south of the city. The Very Large Telescope (VLT) and its support facilities, especially the staff and visitor hotel, were a stunning example of the successful construction in Chile of a truly major astronomical installation. It was encouraging to see, and at the same time provoked a bit of jealousy among the Americans.

One unfortunate turn of events was that it was not possible to rent any Toyota HiLux® 4-wheel-drive trucks from the Hertz rental agency. These short-bed super-cab models, equipped with extra gas, water cans, and spare tires, were favored by the geologists prowling the desert for minerals, as well as by personnel of the mining industry that dominates the region. Instead, only a Chevrolet sport utility vehicle and a more conventional Toyota pickup were available, both more suited to urban use rather than the poorly maintained roads of the desert.

After an overnight stay at the Park Hotel in the mid-sized city of Calama, the team drove to the town of Ollagüe, on the border with Bolivia. Ollagüe is where a railroad line from Antofagasta crosses into Bolivia and on to La Paz, carrying freight and passengers. It is near sites previously visited by the SAO group, and the team was headed to the highest of those sites. The town offered a school, medical clinic, restaurant, and, most significantly for this trip, a police station. Otárola forgot that protocol requires checking in immediately with the Chilean national police, or Carabineros, upon entering a border town. It had been a rough drive on a poor dirt road with the inadequate rental vehicles and the hungry team went to the restaurant first. Upon leaving the restaurant, they were accosted by an officer who asked why they had not checked in with him on arrival. At the station, they presented passports, driver's licenses, and rental agreements. The officer in charge falsely claimed that the truck insurance had expired and that the group would be held at his station, along with the vehicles, until the matter was sorted out. Otárola suspected a bribe was expected. After what seemed like a very long time, another officer entered the station and greeted Otárola warmly. He was a friend. Suddenly, all was in order and the team was free to go. The experience was an insight on the darker side of the tight security procedures that lingered for a while after the dictatorship of Augusto Pinochet had passed.

After visiting the Ollagüe school and medical clinic, the team went to the site that Raffin and Kusonuki had listed as their first choice for the SMA. It was a gentle climb out of the village. The site was spectacular, with a large level of extent. The sky was unbelievably blue. A tipping radiometer was too bulky and complex in its operation for this initial search. Instead, the team used a handheld water vapor meter,[18] provided by Mark Gordon of NRAO Tucson. One pointed it at the Sun to measure the precipitable water vapor (PWV) column in millimeters. The readings indicated very low values, which was encouraging, but the device also seemed erratic and its reliability was questioned. The site near Ollagüe is shown in Figure 4.7.

It had been a long day. The group, tired and reluctant to spend a night in Ollagüe, at its elevation of nearly 5,000 m, decided to drive back to the Park Hotel in Calama. On the way, they were treated to a spectacular view of the full

Figure 4.7 The site identified by SAO's Raffin and Kusonoki as "Ollagüe Area, #19/20" elevation 4,650 m near Cerro Aucanquilcha. It was their top choice for the SMA. Left to right: Hernán Quintana, Angel Otárola, Paul Vanden Bout, and Bob Brown. Credit: Riccardo Giovanelli; NRAO/AUI/NSF, CC BY 3.0.

Moon rising over the Salar Ascotán. The long day, wretched road condition, and high altitude left them with blinding headaches on arrival at the Park Hotel. These were cured by Quintana's therapy of one or more Pisco Sours, a national mixed drink specialty. He suggested making San Pedro de Atacama the next stop. He said the small village was charming, and that he had frequently vacationed there. Furthermore, he knew there was high, flat terrain above the village toward Argentina. As it turned out, Brown also knew of the high, extended terrain above San Pedro based on his recent studies of topographical maps. He had purchased a number of such maps for the general area while in Santiago at the start of the trip. Leo Bronfman had taken him to the Instituto Geográfico Militar, which sold the maps to the public. It was suspected that in border regions the maps had deliberately been made incorrect, but even so, there was obviously high ground at the top of the Jama Pass above and to the east of San Pedro. Most important, Otárola had actually seen the spot on a previous trip. As has been mentioned, in May 1993 with a group from ESO and Japan, he had looked down at the plateau that was to become the MMA site from the south side of Cerro Toco. It was quickly decided to head to San Pedro.

The group arrived in San Pedro late the next day and checked into the Hostería, at that time just about the only decent place to stay in the village other than hostels where one could spread a sleeping bag on the floor. The Hostería had a restaurant and, most significantly, a gas station, the only one in the area. Good use was made of both. The village of 2,500 inhabitants was delightful and enchanting, as Quintana had claimed. The streets were unpaved and electricity was only available between the hours of seven to ten in the evening. Underground runoff from the Andes range provided a stream that flows through the village and drains in a delta, feeding the lagoons of the Salar de Atacama, the third-largest salt flat in the world. San Pedro de Atacama had been a farming village for centuries. A Spanish-built church sits next to the town square. A local priest, Father Gustavo Le Paige, had founded an archeological museum and built a collection of artifacts including well-preserved mummies. But the museum had to wait for later trips; the team was off to the high country east of town.

A national highway, Route 23, runs from Tocopilla, on the Pacific Coast, past the largest open-pit copper mine in the world, Chuquicamata, through Calama to San Pedro and points south. From San Pedro, Route 27 goes over the Jama Pass to Argentina. The team followed this route, past the customs and immigration checkpoint just outside of San Pedro and the military airstrip nearby, up into the Andes. The highway was unpaved at the time and deeply rutted in places as a result of traffic and poor maintenance. About halfway to the top of the pass, the vehicle carrying Vanden Bout and Otárola died. It had a carburetor rather than a fuel injection system, and the carburetor mixture had been tuned to run at sea level. It was surprising that it had not already failed in Ollagüe. Brown, Quintana, and Giovanelli went on to the top of the pass, where they could catch a view of the iconic Laguna Verde in Bolivia. At a high point, an attempt was made to measure the water vapor with the water vapor meter. It would not work, even after installing new batteries and giving it a few sharp raps. But the sky was an astoundingly deep blue and it felt like the area might pass Frank Low's informal test for a good infrared site – a tendency for nose bleeds. Brown chose not to follow Otárola's directions to Cerro Toco in view of the late afternoon hour and concern for the broken vehicle.[19] The team convened again in San Pedro greatly encouraged. In their review of the last two days and the visits to the two high sites, the group clearly favored a site near San Pedro. The next step was to visit the site identified by Otárola on his May 1993 trip. But that would not happen for another five months.

Before returning to the United States, Brown, Giovanelli, and Vanden Bout flew to Mendoza, Argentina, to meet with Raúl Colomb, director of the Argentine Institute of Radioastronomy, and Hugo Levato, the Director of the

Astronomical Complex El Leoncito. They visited the El Leoncito site and spent the night before returning to Santiago and the United States. It was clear that the low elevation of the Argentina site ruled out any possibility of developing the MMA there. In 1995, Felix Mirabel proposed siting the MMA in northern Argentina, near Salta and due east of the Chajnantor site, but the elevation of the area, although higher than El Leoncito, was inferior to Chajnantor for submillimeter observing.

Llano de Chajnantor

The MMA Design and Development Plan that NRAO had submitted to the NSF in 1992 required approvals up through the management structure at NSF, including the National Science Board, in order for the MMA Project to receive funding. Bill Harris, NSF Assistant Director for Mathematics and Physical Sciences (MPS), was key to getting this approval. Informed of the potential site near San Pedro, Dickman thought that Harris might be convinced to promote the MMA by a visit to the Chilean altiplano region. With the stunning vistas as a backdrop, he might warm to the idea of building a millimeter instrument there. On 24 October 1994, Vanden Bout, Harris, and Hugh van Horn, AST Director, arrived in San Pedro. Leo Bronfman joined them. Brown had preceded them to San Pedro. Otárola was unable to guide Brown on this trip as he was needed at SEST for the scheduled maintenance of the telescope. Instead, Geraldo Valladares went with Brown. He had been hired by Bronfman to accompany Otárola on trips to high elevations. Following Otárola's directions Brown and Valladares spent two days exploring the top of the Jama Pass and the side road to Cerro Toco, where Otárola had been able to see the Chajnantor Plateau. They also found a place on Highway 27 to impress the NSF delegation with the surroundings. On 25 October 1994, the group drove up Highway 27 to the spot Brown and Otárola had chosen, near the turnoff to Bolivia and the Laguna Verde. The scenery was as dazzling as expected based on the previous scouting visit – bone dry land, volcano peaks, the green lakes, and endless distant vistas. From the deep blue sky and thin air Harris could see the potential for millimeter observing. Brown told Harris that he was sure that a flat expanse nearby could accommodate the MMA. Harris, in his typical "I want action" style told Brown to find it. On the drive back to San Pedro, Harris ordered van Horn to make the MMA happen, something that van Horn was only too willing to attempt.

The next day Brown and Valladares, following the route they had taken two days earlier, arrived at the lookout point on Cerro Toco and saw the site shown in Figure 4.8. Brown's recollection of seeing the site was given to Vanden Bout in an email message[20]:

Figure 4.8 The 5,000 m (16,400 ft) elevation site as seen from a point on Cerro Toco. M. Gordon named it the Llano de Chajnantor, after a neighboring peak, Cerro Chajnantor. In general usage, the site name is usually shortened to "Chajnantor," a word that means "place of departure" in the Kunza language of the Atacameños. Credit: Simon Radford; NRAO/AUI/NSF, CC BY 3.0.

The next day … I drove with Geraldo Valladares back up the Paso de Jama to find a "real" site for the MMA. It was at that time that Geraldo and I ventured down the road to Bolivia only to be turned around by a guy pointing a rifle at us (Geraldo says 'I do not like his face'). We then drove the opposite way, south, found the western entrance to the plateau by following the switchback road that goes to the sulfur mine on Cerro Toco and continued onto the Chajnantor Plateau. The plateau was perfect. I climbed up on a rock and excitedly took some panoramic pictures with Geraldo watching me perhaps wondering if I had lost my mind. I said to Geraldo, 'Isn't this a beautiful place? Can't you just see it with the MMA antennas spread out in the distance?' Geraldo was very serious. He said, "Yes, but it is a beautiful place without the antennas." … That was October 26, 1994.

Who "discovered" the ALMA site? As with many discoveries, the answer is nuanced. Quintana knew there was flat land at the top of the Jama Pass from family vacations taken in San Pedro and topo maps of the area. Otárola was the first to identify the site on his May 1993 scouting expedition for the LMSA, and he provided Brown with the directions to the Cerro Toco lookout. Brown and Valladares were the first to drive down from Cerro Toco and actually explore the site itself. It is a little like asking who discovered America: the prehistoric people who crossed the Bering Strait to settle the Americas, the Vikings, or Christopher Columbus? A fair statement is that Otárola and the ESO team were the first to *see and identify* the site and that Brown and Valladares were the first science team to *explore* the site.

Chajnantor Site Studies

In November 1994, the NSF National Science Board approved the MMA Design and Development Plan, after a presentation by Dickman. In principle, funding would follow. But, although the Chajnantor Plateau might seem promising as a site for the MMA, NSF expected to see extensive studies to characterize the site before it could be approved as the site of choice. Those studies had just begun.

On 27 October 1994, an NRAO team composed of Peter Napier, Frazer Owen, Simon Radford, and Juan Uson arrived in Chile from Socorro, New Mexico, with a tipping radiometer for measuring atmospheric transparency and the gear necessary to operate it. Their expedition had been organized months earlier by Brown on his return from the first site search in May of that year. On 14 July 1994, Napier had sent Brown an email message[21] listing the places near San Pedro that had sufficiently high elevations to be of interest for transparency measurements with a tipping radiometer, including Chajnantor. By coincidence, they arrived the day after Brown had visited the Chajnantor site. The survey team crossed paths in the Santiago airport with Vanden Bout and the NSF group, who were returning to the United States. Following a discussion with Brown, the arriving team's attention was immediately focused on Chajnantor. After flying to Antofagasta and connecting with Otárola, the site-testing team drove the next day to San Pedro and unpacked the tipper. It was giving problems and working to fix these occupied the following day. On 30 October 1994, the team, guided by Otárola, explored the Chajnantor site looking for a location where they could install the tipper. A boulder was located that offered shelter from the wind and this is where they chose to set up camp. The next day Radford drove Uson, who had become ill, to Antofagasta for his return to the United States. The rest of the team took the tipper to the high site and commenced its operation. This first monitoring station on the Chajnantor site is shown in Figure 4.9.

On 1 November 1994, Otárola left to work with a Japanese team in the area, who were running a radiometer at Paranal. The others stayed in San Pedro where they met a German graduate student working on a meteorology degree. He educated them on the Altiplanic Winter, known colloquially as the "Bolivian Winter," an annual phenomenon in (austral) mid-summer whereby moisture from the Amazon Basin is pushed west over the Andes. In severe events, there can be heavy snowfall on the high plateaus. As the snow melts and runs off the mountains, deep ravines are cut, which are visible everywhere on the mountain slopes. The next day, after checking on the tipper, the group, now Napier, Owen, and Radford, decided to drive to an even more extreme site at an elevation of 5,700 m on the Bolivian border. A dirt road to the site existed by virtue of a sulfur mine, now abandoned. Years later Radford would return to this site and install a submillimeter spectrometer built by Ray Blundell of the SAO. One of the authors (Plunkett)

Figure 4.9 The first MMA site survey team and their equipment: tipping radiometer in the entrance to the tent and solar panels for charging the batteries that powered the tipper in front. Left to right: Peter Napier, Frazer Owen, and Angel Otárola. Credit: Simon Radford; NRAO/AUI/NSF, CC BY 3.0

recalls visiting this site several times while she was a graduate student to make observations using the spectrometer. It was so high and remote that they always drove two trucks and kept both running throughout the night as a precaution.

A total eclipse of the Sun occurred on 3 November 1994. The path of totality passed through the border town of Arica in the far north of Chile, but the team chose to stay in San Pedro, where the event was 92 percent total; their priority was to tend the tipper. After doing so, they decided to climb Cerro Chajnantor, the extinct volcano for which the ALMA site was named by Mark Gordon, as the Llano de Chajnantor. The elevation of 5,635 m proved to be a challenge but Napier and Radford made the summit. The next day the team brought the tipper and equipment back from the Chajnantor Plateau site to San Pedro, and packed up to leave. It was a full day of getting to know the area: a visit to the archeological museum in the village, then a drive of several hundred kilometers to Paranal to compare measurements made there with those of a Japanese instrument,[22] and to Antofagasta for the night. The next day, after picking up the tipper from Paranal, the team flew back to the United States. The tipper measurements made at Paranal agreed with those of the Japanese radiometer to within five percent, so both teams felt their data could be trusted.

For the most part, the PWV measurements recorded on the Chajnantor site were comparable to those on Maunakea. This was encouraging. All the more so, as the readings had been taken at the start of the Altiplanic Winter. Because the tipper had developed some operating problems, it was decided to return it to the United States, overhaul the tipper, and come back and set up a permanent monitoring station.

The Container

Sustained and reliable monitoring of the weather and the atmospheric transparency and stability required a small laboratory that could be deployed on site. Ideally, the laboratory would house test equipment and the tipper, as well as serve as a base to control an instrument to measure the atmosphere stability and record data from the weather station. Gerry Petencin, an engineer on the staff at NRAO based in Socorro, designed and built this laboratory in a shipping container. In April 1995, Petencin, Ramón Gutiérrez, a VLA antenna mechanic, and Otárola installed the container at a location on the Chajnantor site that was thought to be approximately where the center of the MMA would be placed, shown in Figure 4.10. An array of solar panels and a windmill generator supplied power to a bank of batteries. An inverter then supplied AC

Figure 4.10 The atmospheric monitoring laboratory in the center of the Chajnantor site. It was built into a standard shipping container. The array of solar panels leaning on the back side of the container provided power. The tipping mirror mechanism of the 225 GHz radiometer can be seen mounted on the top right-hand side of the container. The small antenna on the end of the container provided a data connection via SatPhone. Credit: Paul Vanden Bout; NRAO/AUI/NSF, CC BY 3.0.

power to the instruments and to the computer that was used to control and record their data. The tipper was mounted inside the container with its tipping antenna poking out through an opening in the wall. It measured the atmospheric transparency at 225 GHz and was later equipped to do so at 325 GHz. The station was brought into operation by Radford, and the tending of the station became the responsibility of Otárola, who was a wizard at fixing problems.

The phase stability of the atmosphere was measured using an interferometer designed by NRAO staff members Darrel Emerson, John Payne, and Dick Thompson, and then installed by Radford. It used two small 1 m diameter antennas separated by about 200 m, to receive the 12 GHz carrier signal from a geostationary satellite. Atmospheric studies on the ALMA site continued for many years. Otárola has curated all the data and results of those studies.[23] In 2005, the temperature variations of atmosphere across the site were measured using radiosonde balloons. The difference of 2 C between two spots 7 km apart was judged to have minimal effect on the pointing of the array.[24] Figure 4.11 shows a balloon launch.

Figure 4.11 Launching a radiosonde balloon during an August 2005 study of the variation of atmospheric temperature across the ALMA site. Left to right: Rubén Bravo (behind balloon), Roberto Rivera, Angel Otárola, Jorge Riquelme. Courtesy of Alison Stirling, reproduced by permission.

The container laboratory, which came to be referred to simply as the "Container," became a destination for those visiting the Chajnantor site. Jim Moran of Harvard University was the first to sign his name on the wall of the Container. He inspired dozens of others to do the same over the years that followed. Figure 4.12 shows a portion of the wall. It is rumored that the wall is now in storage waiting for a suitable place for display.

It must be noted that the effort to measure the atmospheric quality of sites in the Atacama was one of extraordinary cooperation between ESO, NAOJ, and NRAO. Otárola was an employee of ESO who was made available to work with the other two groups by the SEST director, Roy Booth, at the request of Leo Bronfman at U. Chile. The NAOJ was looking for a site for the LMSA and for a single-dish submillimeter wavelength telescope. ESO was searching for a site for their Large Southern Array and NRAO was doing the same for the MMA. All the groups worked together to ensure success, with Otárola at the center. In the end, as will be described in the next chapter, the three projects were merged into one with Chajnantor for its site.

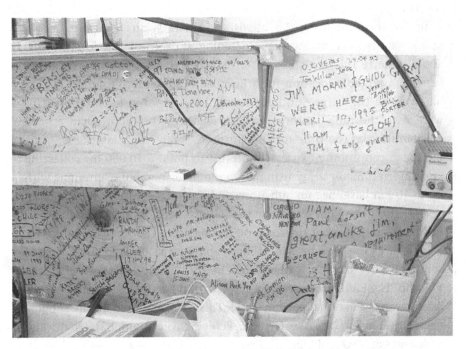

Figure 4.12 A portion of the "signature wall" in the monitoring station, a modified shipping container on the Chajnantor site. Note that on 10 April 1995 when Moran and Guido Garay (U. Chile) visited, the atmospheric opacity was 0.04, or an atmospheric transparency at 225 GHz of 96 percent. Courtesy of James Moran, reproduced by permission.

Searching for the ALMA Site

My involvement in the search for a location in northern Chile for a radio interferometer with a large collecting area and long baselines is one of the highlights of my personal and professional life. I owe a debt of gratitude to my former supervisors and collaborators at ESO for entrusting me to join colleagues from Europe, Japan, and the United States for the exploration of suitable sites in the Atacama Desert. I also contributed to studies to better characterize their conditions in terms of local weather, atmospheric PWV, and atmospheric stability. In all of this work, I was not alone. I was a member of a great team. There were too many to mention, but I must acknowledge the friendship and great collaboration I had with my colleagues Guillermo Delgado and Roberto Rivera, both of whom, sadly, did not live long enough to see the ALMA come to full operation and deliver great science for the scientific community. I think of them often. I am convinced ALMA was the product of a great collaboration that included various organizations with leaders who supported and stimulated all of us to do our best. Believe me, we had a lot of fun working together.

In the beginning of the 1990s, scientists from ESO and Onsala Space Observatory of Sweden, including staff from the Swedish-ESO Submillimetre Telescope (SEST), started to explore the Atacama Desert to identify a possible location for a European radio-interferometer facility (later called the Large Southern Array, LSA). My first exploration trips were with the radio astronomer Lars Bååth, with whom I had a good collaboration at the time for this purpose. A few years later (1995), the European community expressed interest to build a 10,000 m^2 collecting area interferometer array using 15 m diameter parabolic antennas with up to 10 km baselines for sub-arcsecond spatial resolution. We explored several sites: the Vega Valley by the Paranal area, Pampa Loyoques and Pampa Herrera (east of the town of Toconao into the Andes mountains), Pampa El Chino, Pampa San Eulogio and Pampa de Pajonales, located south of Salar de Punta Negra and in the vicinity of the Llullaillaco Volcano. The Llullaillaco area, with its natural beauty and pristine views, later became a national park in Chile. This effort counted on the support from Roy Booth, Lars-Åke Nyman, Peter Shaver, Daniel Hofstadt, and colleagues at OSO and SEST.

Simultaneously, in 1992, a group from the Smithsonian Astrophysical Observatory visited northern Chile, around the west slope of the Andes mountains, to identify potential spaces for the observatory's Submillimeter

Array, which was ultimately built on the Maunakea Science Reserve in Hawaii.

Not long after, an extended exploration campaign along the Atacama Desert was conducted by the Nobeyama Radio Observatory (NRO) of Japan, with help from our local collaboration between the SEST team and academics from the Astronomy Department at the University of Chile. The NRO group wanted to identify a place for their LMSA project. The colleagues from Japan studied in detail two potential sites: Pampa de Río Frío, near the Llullaillaco Volcano, and Pampa La Bola, just east of the Chajnantor Plateau. This exploration effort was led and supported, on the field, by Masato Ishiguro and Naomasa Nakai.

This effort gained major momentum with the arrival in May 1994 of Paul Vanden Bout and Robert Brown from NRAO in the United States. Together with some of their guests and my participation, we visited several sites. They were convinced to come back and look more closely at the high- elevation sites by the Chajnantor Plateau as potential opportunities for their Millimeter Array. Soon after the first visits, we started monitoring the optical depth at 225 GHz and also the atmospheric stability using a 12 GHz, 300 m baseline radio interferometer. In the period from 1996 to 1998, the LMSA and LSA teams also joined NRAO at Chajnantor for a collaborative effort to monitor conditions in the Chajnantor/Pampa La Bola area.

All the teams acknowledged the superior atmospheric qualities of the Chajnantor Plateau to support radio astronomy in the millimeter/submillimeter spectral bands. This opinion was supported by the results from our collaborative studies. The friendly collaboration to identify the best place for what we now call ALMA had a perfect ending. It included people from many countries, organizations, and professional backgrounds and was a great opportunity for my own professional development. My family and I owe a great debt of gratitude to all who made that opportunity possible.

Angel Otárola
European Southern Observatory
Santiago, Chile

Cost Considerations

The Chajnantor site was deemed superior to Maunakea, but was it affordable? To answer this question, Mark Gordon was given the task of studying the operations of the optical observatories in Chile and constructing a strawman construction and operating budget for the MMA. To give Gordon credibility with Chilean contacts, Brown made him Head of MMA Site Development, Chile. After taking an intensive course in Spanish, Gordon made many visits to Chile gathering the relevant data. Using a mathematical estimation technique called "fuzzy logic" he came up with costs. Gordon also investigated means to counter the effects of working at an elevation of 5,000 m; with Eduardo Hardy, he visited facilities in La Paz, Bolivia, where oxygen enrichment was in use. His report,[25] *ALMA in Chile: A Plan for Operations and Site Construction,* is a comprehensive document covering the history of Chile, its geography and geology, climate, and details on the area around the proposed MMA on the Llano de Chajnantor. The report also laid out detailed plans for operations and site construction. Although Chile would be more expensive than Maunakea as an MMA site, the premium was considered good value. These plans were later captured in the chapters he wrote for the MMA Project Book on site development and operations. The report was an important input to the decision by NRAO management to place the MMA in Chile. After the MMA entered ALMA, a much larger project, the site development and operations costs increased accordingly, but Gordon's analysis of the issues and his advice provided valuable guidance beyond the MMA.

Site Proposal

As the data from the NRAO container came in and were analyzed, it became obvious that the site was the best that had been studied. The MMA project came to assume that it was the site of choice. Dickman, NRAO's NSF program officer, was keenly aware of NSF requirements and told NRAO it would need to follow proper protocol. He asked for an official proposal requesting approval of Chajnantor as the MMA site. The first draft of the site proposal was available for comment on 9 October 1996. In 33 pages, the report[26] presented the science requirements, the site testing program, and the two candidate sites, Maunakea and the Llano de Chajnantor. The characteristics of the two candidate sites were given, concluding that the Chile site was best. The report went on to argue the feasibility of locating the MMA on the Chajnantor site with respect to volcanism, earthquakes, high altitude considerations, and logistics. The issue of lightning on the site was not mentioned, even though it was potentially a serious threat. (Six years later a study[27] showed that lightning was rarely observed on Chajnantor.) Nor was the matter of exposure to cosmic ray radiation discussed. (Five years later a study[28] showed that the radiation dosage for an ALMA employee splitting his/her time between the high and mid-level ALMA sites was approximately double that at sea level.) Strawman plans for construction, commissioning, and operations were presented. The conclusion of the report was that Chajnantor was the site with the best atmosphere, could accommodate the longest anticipated separations of antennas, and was operationally feasible.

The data supporting the selection of the Llano de Chajnantor for the MMA were clear. Figure 4.13 shows how a potential array 3 km in extent could be accommodated on it and Maunakea. But it is obvious that Chajnantor offers a much large flat area with the possibility of much larger arrays, while on Maunakea the 3 km array barely fits on sloping terrain. Figure 4.14 shows the high-frequency transparency on Maunakea and the Chajnantor site.

The proposal to site the MMA in Chile was not officially submitted to the NSF until 18 May 1998. The cover letter to Dickman, signed by Brown, was accompanied by an endorsement from the AUI Board, and signed by Martha Haynes, Interim AUI President. On 31 December 1998, Aaron Asrael, the Grants and Agreements Officer for the NSF, sent Haynes a letter approving the choice of the Llano de Chajnantor as the site for the construction of the MMA. He took pains to point out that actual construction would require the approval of the National Science Board.

Protecting and Securing the Site

Mining has been the principal, and at times only, industry besides tourism in northern Chile, and it has a long, important legacy for the country's identity and economy. It began with the extraction of saltpeter (sodium nitrate, colloquially known as "white gold"), but it is copper that came to dominate mining in modern-day Chile, with lithium recently rising in importance. The Chuquicamata mine is the largest open-pit copper mine in the world, towering above Calama from the north. Another major mining operation, Mina Escondida, is located midway between the Pacific coast and the southern end of the Salar de Atacama. Other mines pepper the region's higher elevations, and geologists are often seen scouring the areas in between for mineral deposits. Even a tourist can understand why they come here, based on the infinite shades that make up any rock formation in sight. The region seems geologically unparalleled.

The economic importance of mining to the Chilean economy is enormous and mining claims take legal precedence over land ownership. It was vital that the MMA project protect the Llano de Chajnantor, which was

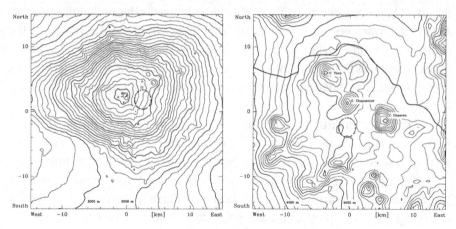

Figure 4.13 Contour maps of the two candidate sites, Maunakea (left panel) and Chajnantor (right panel). A small dotted oval in each map shows the possible locations and extents of a nominal 3 km array. The Chajnantor site clearly offers much more level ground for more extended baselines. Credits: Left panel –Adapted from a topographic map of the US Geological Survey; NRAO/AUI/NSF, CC BY 3.0; Right panel –Adapted from a topographic map of the Instituto Geográfico Militar, NRAO/AUI/NSF, CC BY 3.0.

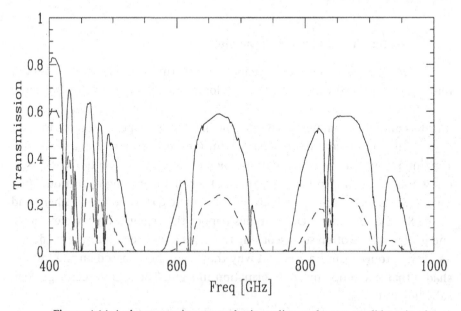

Figure 4.14 A plot comparing atmospheric quality on the two candidate sites in the submillimeter band. The dark line shows that Chajnantor has two to three times better transparency than Maunakea (dashed line) for the best observing conditions. Credit: Simon Radford; NRAO/AUI/NSF, CC BY 3.0.

government land held by the Ministerio de Bienes Nacionales (BN), from mining interests. The array could not be built there if there was even the slightest chance the site would become an open-pit mine. The question was: had any mining claims been filed on the site? If not, how could the MMA project file a claim before anyone else? As rumors of the MMA project spread, there was a danger that some party would file a mining claim on the site, if only to demand a ransom to relinquish the claim.

NRAO was assisted in filing the mining claim by Leo Bronfman. To build and operate an astronomical observatory in Chile requires an agreement with U. Chile regarding the sharing of observing time. This is a legal requirement, under Chilean Law 15172. (ESO, as an international treaty organization, did not need to satisfy this law but rather dealt directly with the Ministry of Foreign Affairs.) It was clear that NRAO would follow the conventional route with the University, and a good working relationship was established in time with the National Astronomical Observatory of U. Chile. The first AUI offices in Chile were located at the observatory on Cerro Calán in Santiago, where the U. Chile astronomy department was also located. Bronfman put NRAO in touch with an attorney who could take care of the necessary land rights.

From the Southern Mini to ALMA

Millimeter wave astronomy in Chile began with the 1.2 m Southern Millimeter-wave Telescope, known as the "Southern Mini," which was brought in 1982 to CTIO by Columbia University and Universidad de Chile (U. Chile), to complete the first CO survey of the Galaxy. The telescope demonstrated the high quality of the atmosphere for mm observing, further demonstrated in the sub-mm by the SEST 15 m telescope installed at La Silla Observatory in 1986/87 by ESO and OSO.

By 1993/94 Europe, North America, and East Asia were working on separate arrays, searching for sites near San Pedro de Atacama. NAOJ characterized several sites, with U. Chile help. The data obtained near Cerro Toco were shared with NRAO. In October 1994, NRAO and U. Chile personnel explored the nearby Chajnantor Plateau, which became the future ALMA site. In 1995 and 1996, permits were obtained from Ministerio de Bienes Nationales (BN) by U. Chile to test the Chajnantor site and Pampa La Bola. In 1998, the whole area was protected for astronomy by CONICYT as a Science Preserve, in a five-year concession from BN. In 1997 and 2001,

AUI/NRAO, and in 2004 NAOJ, signed collaboration agreements with U. Chile, officializing their status as international organizations, allowing 10 percent of its future observing time in ALMA for Chilean astronomy. Similarly, ESO signed an ALMA agreement with the Chilean government in 2002.

The transition to a long-term BN concession for the ALMA consortium, in 2003, was marked by an accord between ALMA and the Chilean government establishing an ALMA-CONICYT fund for the development of Chilean astronomy, and an ALMA Region II fund promoting social and economic development in the local community. The area surrounding the ALMA site, including Cerro Chajnantor, has excellent mid-infrared conditions. It was protected by CONICYT in 2005 by the creation of the "Atacama Astronomical Park." After a long-term concession from BN was given, the park was inaugurated in 2014 further shielding ALMA.

Together with the progress of submillimeter astronomy in Chile, an important consequence of the ALMA installation was the opportunity to participate in state-of-the-art astronomical instrumentation projects. The U. Chile Millimeter-wave Laboratory (MWL) obtained funding from CONICYT to develop in 2008–2013 an ALMA Band 1 receiver prototype, within the new Basal Center for Astrophysics and Associated Technologies, with support from ALMA and Caltech. The MWL was staffed by PhD students and engineers, some trained at the Caltech Cosmic Background Interferometer. Collaborations were further established with NRC/HIAA (Canada), ASIAA (Taiwan), NRAO, U. Manchester (UK), and the Centro de Astronómico de Yebes (Spain). The MWL participated in the development of six ALMA Band 5 prototype receivers, within an ESO-led consortium. The Southern Mini was relocated from CTIO to Cerro Calán Observatory in 2010, close to the MWL. The Band 1 optics design (horns and lenses) was further optimized and prototyped in 2014–2018, in collaboration with NAOJ. Production for the ALMA Band 1 receiver suite took place at the MWL in 2019–2021. Work with NAOJ is underway for the optics of the new ALMA Band 2. Having ALMA in Chile has been crucial for the establishment of new technological capabilities, and a great example of international partnership. Personally, it has been a privilege to be involved.

Leonardo Bronfman,
Professor, Universidad de Chile
Santiago, Chile

The lawyer, Carlos Ruíz Bourgeois, had a distinguished career with clients in the mining industry. He had been a professor at the U. Chile law school and had held high-level administrative positions there. He was at the time retired, but still consulting on matters of mining law. His son, Juan E. Ruíz was also an attorney, working for the Center for Space Studies at U. Chile. All of the dealings with Ruíz senior were conducted through the son. It was clear that the well-connected father would pave the way for the son, who would do all the work. Brown and James Desmond, the NRAO Associate Director for Administration, were given powers of attorney to represent AUI in Chile. The powers of attorney were crafted after those used by AURA and the Carnegie Institution of Washington (CIW). In July/August 1996, a contract for the legal services of C. Ruíz was signed. The mining claim to be filed covered 20,050 hectares (almost 50,000 acres), the largest mining claim in the history of Chile. It would cost AUI approximately $50,000 in filing fees and approximately $25,000 in annual patent fees, plus the fees and expenses of the Ruíz lawyers. On 14 October 1996, notice of the claim was published in the official Chilean mining periodical, the Boletín Oficial de Minería, Volume 26.508, pp. 76–79. The fees were due in 30 days. After paying the fees, Brown made a trip to Santiago where J. Ruíz transferred the claim, which had been filed in his name, to Brown on behalf of AUI.

Legal Status in Chile for AUI

From this point on, AUI's affairs in Chile were handled by Eduardo Hardy, who was appointed Gerente General (General Manager) for AUI in Chile. He had been recruited on the suggestion of Riccardo Giovanelli when he learned that NRAO was seeking a representative to conduct MMA affairs in Santiago. Hardy was fluent in Spanish, having grown up in Santiago from the age of 10, and was fluent in English and French as well. He was a research astronomer with a degree from the University of Indiana; his doctoral thesis had been supervised by Alan Sandage of the Carnegie Observatories. At the time he was recruited by AUI, he was a tenured professor at Laval University in Quebec City, Canada. But he was apparently ready for a change, and returning to Chile was an attractive prospect. His appointment became official on 1 January 1998. He would prove to be a significant asset in future negotiations with the Chilean government at all levels.

In particular, in the early days of NRAO/AUI activities in Chile, Hardy was the point of contact in securing legal status for AUI in Chile. The procedure for gaining legal status was the same as that followed by AURA and the CIW decades earlier, so at least there was precedent. To begin, a cooperative

Figure 4.15 Left to right: Eduardo Hardy, President Eduardo Frei, Martha Haynes, who was interim AUI president at the time. The framed map was constructed from satellite images by a group at Cornell University. Credit: Paul Vanden Bout; NRAO/AUI/NSF, CC BY 3.0.

agreement was concluded in 1997, granting the astronomers at U. Chile 10 percent of the observing time on the MMA, in accordance with the Chilean Law 15172. Recognizing the agreement, the Ministry of Foreign Affairs asked President Eduardo Frei to issue a decree granting AUI the same rights and privileges as ESO, and conceding the site to the Comisión Nacional de Investigación Scientifica y Tecnológica (CONICYT). That occurred on 15 June 1998. In recognition of this decree, in 1999, the Chilean Treasury Department issued a decree granting AUI exemption from import duty and value-added taxes. (In 2002, both the agreement with U. Chile and the Treasury Department decree were amended, changing the name MMA to ALMA.) At a press conference on 21 August 1998, President Frei declared the area around the site as the Chajnantor Scientific Preserve, which gave it a further degree of protection from interfering activities such as mining. The declaration was also signed by Sergio Jiménez (Minister of Mining), Adriana Delpiano (Minister of National Assets), and Mauricio Sarrazin (President of CONICYT). The declaration was welcome, but AUI continued to maintain its mining claims, as a precaution in case the designation could be reversed by a later president, however unlikely a prospect that might have been. Figure 4.15 is a photograph taken at the press conference for signing the declaration. On 7 July 1999, NRAO/AUI received permission from CONICYT to conduct *official* site studies on Chajnantor. ESO's permission came on 25 January 2000 and NAOJ's in July 2001, for the LSA and LMSA, respectively.

Radio Frequency Interference

The Chajnantor site needed more than protection from mining interests. It needed protection from interference from radio transmitters in the area. What the MMA needed was a so-called "quiet zone," such as those in place for the NRAO Green Bank, West Virginia, site and for the Arecibo Observatory. The Telecommunications Subsecretary of Chile, known as SUBTEL, created a quiet zone of radius 30 km around the center of the site, specifically, for frequencies in the range 31–275 GHz. In addition, it created a coordination zone of radius 120 km, within which applications for the installation of a fixed location transmitter at these frequencies was subject to review. The quiet zone and its local protections do not have any official *regulatory* recognition at the International Telecommunication Union Radiocommunication Sector (ITU-R) but the telescope's frequency use was recorded in the Master International Frequency Register. The relevant decrees were issued in 2003 and 2004, and the ALMA site was registered with the ITU-R. The team that gained this protection for the MMA included Benjamin Jacard (U. Chile), Harvey Liszt (NRAO), Darrel Emerson (NRAO), and Tom Gergely (NSF). The effects of vehicular radar from traffic on the highway over the Jama Pass, a major connection between Chile and Argentina, remained a concern. Nothing could be done to stop potential interference from satellites. Their transmissions had not yet reached millimeter wavelengths, but their centimeter band transmissions could potentially interfere with the MMA's electronics.

NRAO had found and secured a site for the MMA, protected it from intrusion by the mining industry, and obtained a degree of protection for the site from radio frequency interference. But the proposal to actually design, develop, and build the MMA was still waiting to be funded. In the next chapter, we tell the story of how MMA funding became a reality as the result of a breakthrough, one that would make it an international project eventually called ALMA.

Notes

1 Owen, F. 1982, *The Concept of a Millimeter Array*, MMA Memo #1. https://library.nrao.edu/public/memos/alma/main/memo001.pdf.

2 McKinnon, M. June 1987, *Measurement of Atmospheric Opacity Due to Water Vapor at 225 GHz*, MMA Memo #40. https://library.nrao.edu/public/memos/alma/main/memo040.pdf.

3 Wade, C. July 1993, *Search for Possible Millimeter Array Sites on the U.S. Mainland*, NAA-NRAO, MMA, MMA Site Selection, Box 1.

4 Warnock, G.,18 November 1991, NAA-NRAO, MMA, MMA Site Selection, Box 1.

5 Napier to Brown, 16 May 1991, NAA-NRAO, MMA, MMA Site Selection, Box 1.

6 NAA-NRAO, MMA, MMA Bahcall Site Selection, Box 1.

7 Liu, Z.-Y. 1987, *225 GHz Atmospheric User's Manual* (Electronics Division Internal Memo #271) MMA Memo #41. https://library.nrao.edu/public/memos/alma/main/memo041.pdf.

8 For the VLBA antenna on Maunakea NRAO agreed to give IfA 15 percent of the *single-dish* observing time. IfA had no intention of requesting such time, but the agreement avoided breaking the rule.

9 An early description of the LMSA was presented by Ishiguro (1997), of European plans for a large mm array by Booth (1997), and of the MMA by Brown (1997).

10 At high elevations rain and snow occur more frequently than on the desert floor. The evidence for this is clear in the many ravines that run down the mountainsides, created by erosion from snowmelt runoff.

11 Nearly a decade prior to Gordon advocating Chile as the place to site the MMA, Ken Kellermann had suggested to NRAO director Heeschen that were a millimeter array to be built by NRAO, it should be put in Bolivia. This idea came to him on a trip via taxi from Lima, Peru, to La Paz, Bolivia, where he saw the vast, flat expanses of high-altitude land. The trip was an adventure shared with Ron Ekers and Radakrishnan following the 18th General Assembly of URSI in 1975. (Kellermann to Vanden Bout, private communication.) Much later, George Wallerstein made the same suggestion to Vanden Bout in a letter but he noted that Bolivia did not enjoy stable government.

12 A report of their searches was given by Otárola, Delgado, and Bååth (1995).

13 The survey team included Naomasa Nakai (NAOJ), Kimiaki Kawara (NAOJ), Nagayoshi Ohashi (NAOJ), Nick Whyborn (ESO), and Wolfgang Wild (ESO), in addition to the team leader Angel Otárola (SEST/ESO).

14 Records of the LMSA site searches have been published by Kono et al. (1995) and Sakamoto (2001).

15 Raffin, Phillippe and Kusonuki, A, *Searching for Submm Sites in Chile*, [SAO] Submillimeter Array Technical Memorandum 59, 14 May 1992. NAA-RLB, Box 2. https://science.nrao.edu/about/publications/alma.

16 Illness forced Nakai and Wild to remain in San Pedro. (Otárola to Vanden Bout, private communication.)

17 Whyborn to Vanden Bout, private communication.

18 The water vapor meter, more properly, an infrared hygrometer, has an interesting history that has been described by Gordon (2005), p. 177.

19 Quintana to Vanden Bout, private communication.

20 Bob Brown *Reminiscence of 1994 Chile Trip*, Brown to Vanden Bout (undated, ca. 2004). NAA-RLB, MMA. https://science.nrao.edu/about/publications/alma.

21 Napier to Brown, NAA-NRAO, MMA, MMA Site Selection, Box 1.

22 The Japanese radiometer was being prepared by Naomasa Nakai (Kwansei Gakuin University) for measurements of atmospheric transparency at Río Frío, a site being considered for the LMSA.

23 The data and results of the Chajnantor atmospheric studies can be found at:
    https://science.nrao.edu/facilities/alma/site-characterization; https://alma.sc.eso.org/;
    https://researchers.alma-telescope.jp/e/report/site/.

24 ALMA Memo No. 541 – *Horizontal Temperature Variations at Chajnantor*. https://library
    .nrao.edu/public/memos/alma/main/memo541.pdf.

25 Gordon's final report, written for the record, is dated 30 May 2000 (revised 3 August
    2000) but the report material was available to NRAO management much earlier in a
    series of draft reports sent to Brown. The final report can be found at NAA, Papers of
    Mark A. Gordon, ALMA, Box 2. https://science.nrao.edu/about/publications/alma.

26 *Recommended Site for the MMA*, May 1998 NAA-NRAO, MMA, MMA Site Selection. https://
    science.nrao.edu/about/publications/alma.

27 ALMA Memo No. 487 – *Lightning Near Cerro Chascón*. https://library.nrao.edu/public/memos/
    alma/main/memo487.pdf.

28 ALMA Memo No. 466 – *Levels of radiation exposure near AOS and OSF*. https://library.nrao.edu/
    public/memos/alma/main/memo446.pdf.

5

Foreign Affairs

"It is our true policy to steer clear of permanent alliance
with any portion of the foreign world."

George Washington, farewell address, 1796

Funding Hurdles

The path was long for the MMA as it progressed from the time of the
proposal submission to NSF, through the subsequent review, en route to being
funded. Myriad hurdles had to be overcome. None could assure success; any
could lead to failure. Several of the gatekeepers were committees that met on
a regular schedule, for example, the NRAO Users and Visiting Committees, the
ACAST, and the MPS Advisory Committee. A strong endorsement of the MMA
in each of their reports was necessary to sustain the momentum of the project. While the NRAO committees, composed of radio astronomers, could be
counted on for enthusiasm, most of the ACAST were in other fields of astronomy and needed to be convinced of the power of the MMA to address questions
that they themselves regarded as important. Luckily, a demonstration of that
power became available.

In the summer of 1991, Bob Brown was intrigued by a preprint of a scientific article received by the NRAO Library. The authors had identified an
infrared source, IRAS F+10214, with a galaxy at a redshift of 2.3 (more than
30 billion light-years away). An infrared source detected at this distance was
sure to be a highly luminous infrared galaxy. In turn, infrared emission indicates the presence of star formation regions in molecular clouds of dust and
gas. Brown wanted to search for the molecular gas. At first, he considered
searching for the lowest frequency spectral line of carbon monoxide using

the 140 Foot Telescope in Green Bank, where he had previously searched for highly redshifted atomic hydrogen. Instead, he agreed to Vanden Bout's suggestion that they look for a higher frequency CO spectral line using the 12 Meter Telescope. Despite the poor observing conditions – it was summer, the rainy season on Kitt Peak – they detected the targeted molecular line. The discovery paper[1] they published was flawed. Systematic errors in the data made the line appear stronger than it actually was, but the detection was real. Furthermore, it was at a redshift more than 10 times larger than any previous CO detection in an external galaxy. When the spectrum was shown to the ACAST as part of a report on MMA progress, the members immediately grasped the potential of millimeter wave facilities for the study of star formation in the early Universe.

With a strong endorsement from the ACAST, Hugh van Horn, Director of the NSF Division of Astronomical Sciences, could speak forcefully at the NSF in support of the MMA project. On their exploratory trip to the high area above San Pedro de Atacama, Harris and van Horn had been awed by the potential for astronomical observations afforded by clear, dry skies of the high-altitude plateau. Harris became a strong advocate within the NSF for the MMA. As was noted previously, the MMA had received a high ranking for major ground-based projects from the report of the 1990 Decadal Survey of Astronomy and Astrophysics. The MMA was ranked second, behind the Gemini North optical/infrared telescope, near the top of the list for new, large, NSF projects. Gemini was organized as an international partnership, portending the next hurdle for the MMA.

International Participation

Organizing large NSF projects as international partnerships became fashionable while Bloch was the NSF Director. He had attended a meeting in Europe organized by the heads of European science funding agencies to discuss the benefits of international participation in future large projects. One perceived benefit was cost savings. The Gemini Observatory is an early example of an NSF international project in astronomy. Undoubtedly, the sharing of construction costs among foreign partners appealed to the US Congress as they reviewed NSF's budget requests for Gemini. That initial funding success made it clear that in the future all large projects proposed to NSF would be wise to include international partners.

Faithful to George Washington's dictum, NRAO had long avoided international cooperation in telescope operations. The sole exception involved the quest for the longest possible baselines for Very Long Baseline

Interferometry (VLBI) observations. These had to span the globe and inevitably led to joint operations with organizations in other countries. While welcoming qualified observers from other countries to use its telescopes, the construction and operation of those telescopes was strictly held as NRAO's sole responsibility. When the 25 Meter project was struggling for funding, Peter Mezger, a director at the Max Planck Institut für Radioastronomie (MPIfR), offered to partner in a 25 m telescope on Maunakea in Hawaii. The NRAO director at the time, Dave Heeschen, declined the German partnership. During the construction of the VLBA, Canada offered its Dominion Radio Astrophysical Observatory site as the location for an antenna. NRAO demurred, placing it nearby in northeast Washington State. To overcome this tradition of independence required direct orders from NSF.

That order came while Neal Lane was NSF Director. Lane had been heavily lobbied to fund the MMA proposal in an effort organized by John Bahcall, chairman of the 1990 Decadal Survey. The MMA was Bahcall's personal top priority for a new large ground-based facility, and he had asked colleagues who knew either Lane or members of the National Science Board to write and urge funding for the project. The pressure reached the point where NRAO, through AST, was told to "*call off the dogs.*" But Lane did respond. On 13 May 1996, he invited Vanden Bout to a breakfast meeting away from his NSF office. He said he intended to include design and development funding in the NSF budget request but, importantly, insisted that the project had to be international. Lane shared Bloch's belief that international projects sold better in the US Congress. A person present pointed out the obvious – successful funding would depend on a champion of the MMA in the Congress. Typically, that would be a senator from the state where the project was to be built. "*Who is the Senator from Chile?*" he asked. Vanden Bout opined that the champion could be Senator Pietro V. ("Pete") Domenici of New Mexico. In fact, Domenici's help turned out to be critical.

Soon after that meeting, Lane had the opportunity to seek support from NASA for the MMA when he found himself seated at a dinner next to Dan Goldin, the NASA Administrator. He mentioned the MMA and its potential for detecting planet formation around nearby stars. Goldin had not heard of the MMA, but was intrigued and wanted to learn more. As a result, Vanden Bout visited Charles Pellerin, then head of NASA's Astronomical Science Division, to sell the MMA as part of NASA's Origins Program. The visit did not go well. Pellerin had established a hard-and-fast rule: NASA does space projects and NSF does ground-based projects. To emphasize the point, he handed Vanden Bout a Washington DC Metro ticket and said, "*I think NSF is near the Blue Line Ballston Station.*"

Selling the MMA Abroad

NRAO's first approach to an international collaboration for the MMA was to seek minority partners in Europe, specifically, the Netherlands. Vanden Bout made a trip to the Netherlands on 13–19 February 1995. Visits were scheduled for talks at the Universities of Groningen and Leiden. Vanden Bout and Ewine van Dishoeck were accompanied by Anneila Sargent of Caltech, who addressed potential participation of the Netherlands in the OVRO and BIMA arrays. Immediately prior to the talks, the Netherlands Committee on Astronomy met to discuss joining the MMA. They concluded that negotiations should begin with possible millimeter astronomy partners and that a proposal should be prepared at the level of 30 million Dfl for Dutch participation. Professor van Dishoeck, a prominent astrochemist and millimeter astronomy enthusiast at the University of Leiden, led the writing of a proposal[2] to the Netherlands Research Council (NWO) requesting funding for a 10 percent partnership in the MMA. The amount requested was $20 million over three years beginning in 1997.

NWO sent the proposal to five referees, one each in the United States, Germany, France, Japan, and England. Their reports[3] were uniformly positive regarding the scientific goals but mixed on whether joining the MMA was the best way to realize those goals. The referee from France wrote a long report that pointed out the shortcomings of the MMA, which was too small in his opinion, and the advantages of joining an all-European effort to build a much larger array, the Large Southern Array (LSA) (to be discussed shortly). At the same time, prompted by IRAM, the Secretary General of the Max Planck Society wrote to Reinder van Duinen, head of NWO, urging consideration of a partnership with IRAM on a large millimeter array. The letter[4] contained the prescient statement: *"Noting that there is also interest in the Japanese radioastronomical community to build a large millimeter array in the southern hemisphere, several future scenarios can be envisaged, including one of truly international collaboration in which the MMA, the LSA, and the Japanese project would all be co-located."* All this gave NWO pause and on 17 April 1996, the proposers were informed by Harvey Butcher, head of the Netherlands Foundation for Radio Astronomy (NFRA), that the proposal had been declined.

Amidst discussions between NRAO and European partners to explore a merger of the MMA and LSA, Canada was approached as a potential US partner. After many presentations, discussions, and recommendations from advisory groups, an agreement with Canada, called the North American Program in Radio Astronomy (NAPRA), a play on the acronym for the North American Free Trade Agreement (NAFTA), was signed in September of 2001 between NRAO and the Herzberg Institute of Astronomy and Astrophysics (HIAA) and the National Research Council (NRC) of Canada whereby the Dominion Radio Astrophysical Observatory would construct a new signal correlator for NRAO's Very Large Array in exchange for considering Canadian applicants for NRAO observing time on the same basis as US applicants, including time on the MMA-LSA merger, by then called ALMA. NRAO had always in its Open Skies policy considered qualified applicants for observing time equally no matter their country or institutional affiliation. But the possibility that this might change if the project partnership turned international was an incentive to join NAPRA; the division of observing time on existing international observatories was intended to be in proportion to the partners contributions to construction and operation. In time, Open Skies became an NSF policy for all its astronomical facilities, international or not, and the NAPRA agreement became moot. The NAPRA agreement expired in 2011, before ALMA became operational.

In fact, Canada did join the United States in the ALMA project. Prior to the NAPRA Agreement, a letter of intent to do so was signed by Arthur Carty, President of the NRC of Canada, and Rita Colwell, NSF Director. That intent was realized in June 2003 with the signing by these parties of an agreement whereby Canada would contribute about 10 percent of the North American obligation to ALMA construction. US and Canadian observing time was to be pooled. Further agreements between NRAO and HIAA/University of Calgary defined the structure of the partnership and the deliverables.

Europe's Large Southern Array

As has been alluded to throughout, the NRAO was not the only group to realize that a millimeter wave interferometer should be the next major astronomical facility to answer the pressing scientific questions of the era. We'll take our narrative back a few years in order to track the activity of the European radio astronomers at the time. The concept that led to the Large Southern Array (LSA) was the brainchild of Roy Booth, director of the Onsala Space Observatory (OSO) in Sweden. In response to Booth's proposal to establish the Sweden-ESO Submillimetre Telescope (SEST), Lo Woltjer, ESO Director General at that time, agreed to a partnership with OSO and to place the new single-dish telescope at ESO's La Silla site. Peter Shaver, ESO staff scientist, was appointed to be the liaison for the project. SEST operated from 1987 to 2003, and its output clearly demonstrated the scientific potential of millimeter/submillimeter observations in the southern hemisphere. The telescope was a copy of the antennas used in the IRAM interferometer on the Plateau de Bure, near Grenoble, France. Booth's concept, the so-called Southern Millimetre Array, was more or less a southern hemisphere version of the IRAM interferometer: (10)

8 m diameter antennas were to be located in what he called Millimeter Valley below ESO's Paranal Observatory. He presented this concept at the end of the second SEST Users Meeting at ESO Garching, 22–23 May 1991. Bob Brown was invited to the meeting[5] and presented the plans for the MMA.

In December 1991, the first meeting of the Southern Millimetre Array Working Group[6] (SMAWG) was held at IRAM. It had 10 members and was chaired by Booth. Arriving late, following some discussion of the science case and array concepts, IRAM staff scientist Dennis Downes presented a compelling argument for what became the defining characteristic of the LSA. He knew the story of IRAS F+10214 and pointed out that Brown and Vanden Bout's detection of CO at a redshift of 2.3 effectively opened the high-redshift Universe to millimeter observations. In a later conference presentation,[7] Downes noted that, "... *for the first time, radio astronomy can study thermal dust emission and thermal line emission from cool, neutral molecular matter at epochs close to the formation of galaxies, ...*" However, IRAS F+10214 was gravitationally lensed, that is, the signal was focused and magnified by a factor of order 10 by an intervening galaxy along the line of sight. Such systems were certainly rarer than un-lensed high-redshift galaxies. To study star formation in the large, un-lensed population would require an array with a huge collecting area, 7,000–10,000 m^2, something like (40) 15 m diameter antennas. From then on, the European millimeter array was to have a large collecting area. Hence, its name became Large Southern Array.

Following this meeting, a search was begun in Chile for a suitable site. Since the LSA science goals were focused on using the CO transitions at 115 and 230 GHz and not at higher frequencies, the search could be satisfied by an area offering 10 km antenna spacings (about 1 arcsecond resolution) at the elevation of the salt flats south of San Pedro de Atacama. Site studies were led by Angel Otárola, in partnership with a Japanese team who were also considering a site in Chile for their project, the LMSA. As was discussed in the previous chapter, the two sites of principle interest to the LSA were Pampa El Chino at 3,300 m and Pampa San Eulogio at 3,750 m elevation, respectively. Both lay side by side at the southern end of the Salar de Punta Negra, 80 km east of the Escondida copper mine.

The European millimeter astronomers continued to have meetings that discussed plans for the LSA and searched for means to get it built. The underlying assumption was that IRAM would lead the project. But the IRAM director, Michael Grewing, was never presented with the right circumstances to seize the opportunity. Meanwhile, the current Director General of ESO, Riccardo Giacconi faced two tasks: finishing construction of the VLT and finding the means to operate it within ESO's budget. The latter meant scaling back or even closing the La Silla site, ESO's first observatory. He asked for reviews of all activities there, including SEST. It is ironic that Giacconi's wish possibly to close SEST would lead to ESO's participation in ALMA. Shaver was charged with producing a justification of SEST, and quickly rallied the community to produce a document of 120 pages presenting its science accomplishments at millimeter wavelengths and concluding with a proposal that ESO provide support for an LSA. On 5–6 May 1994, that document,[8] entitled *(Sub)Millimetre Astronomy at ESO*, was favorably reviewed by the ESO Scientific Technical Committee (STC). The STC was chaired by Steve Beckwith, a director of the Max Planck Institut für Astronomie (MPIA) in Heidelberg, Germany, an infrared astronomer interested in star and planet formation. The STC endorsed the five points[9] made by the SMAWG: (1–3) The productivity of SEST, the recommendation for continued support of SEST, and the need for development of instrumentation for SEST; (4) the conclusion that the future of millimeter wave astronomy lay in large arrays; and (5) the need for a permanent millimeter advisory group. By October 1996, all these points had been fulfilled.

The first meeting of the ESO Millimetre Advisory Group was held in August 1994. It had 10 members, among them Booth, Grewing, Shaver, and Butcher. It recommended preparation of a concept document, starting a design study, and holding a workshop on relevant topics. In April 1995, a memorandum of understanding (MOU)[10] was concluded between OSO, ESO, IRAM, and NFRA to pool resources for a study of a large millimeter array in the southern hemisphere and prepare a report within two years. Downes wrote a brochure that presented the concept, published in October 1995. The feasibility studies were then organized. And on 11–13 December 1995, Shaver organized a workshop that was held at ESO's headquarters in Garching, Garmany, on science with large millimeter arrays. The workshop hosted 95 participants, overwhelmingly European, but including attendees from Australia, Canada, Chile, Japan, Mexico, and the United States. Among the many contributions, Downes presented a detailed description of the planned LSA. This was followed by examples of extragalactic and cosmological studies, as well as galactic and solar system studies. James Lequeux discussed the synergy between the LSA and the VLT. The concluding remarks were made by Lo Woltjer. He summarized the case[11] powerfully: *"The scientific case for such an array is overwhelming ... a perfect counterpart to the Hubble Space Telescope with comparable resolution but unhindered by dust opacity ... highly complementary to the VLT."*

At its 30–31 October 1996 meeting in Milan, the STC reviewed the progress that had been made and endorsed the LSA as a possible ESO project. Beckwith recalls[12] the STC initially being concerned that the LSA would compromise the plans for the OverWhelmingly Large Telescope (OWL), a 100 m diameter optical telescope proposed as ESO's next step after the VLT. When it was realized that this behemoth lay many years in the future, the committee recognized not only the scientific potential of the LSA but also that it could be a budgetary bridge between the VLT and OWL. The STC report favorably impressed Giacconi although he was heard to remark,[13] "*Let's not go overboard.*"

The Large Southern Array

The European initiative for a large millimeter array arose rapidly in the mid-1990s, and it was a very exciting time. A number of unrelated developments remarkably coincided to make such a facility possible.

By the early 1990s, European millimeter astronomy was quite well established. Sweden and ESO operated a 15-Meter Telescope (SEST) in Chile, and IRAM operated a 30-Meter Telescope in Spain and an array of four 15-meter antennas in France. The idea of building a similar array in Chile, proposed by Roy Booth, was under discussion. But it was drastically revised by the startling discovery at NRAO of CO line emission from a $z = 2.3$ galaxy. Suddenly it was realized that we could potentially see the earliest galaxies in the universe using a very large array, one with a total collecting area of 10,000 m^2.

But how could such a giant project be funded and operated? In early 1994, ESO's Director General was looking to provide funds for VLT operations, and SEST was in his sights. He asked for a report, and got a surprise: not only was SEST strongly supported, ESO was also asked to provide support for a huge new millimeter array. His later response to the enthusiasm: "*This could be ESO's next major project!*"

There followed a few hectic years of scientific and technical studies. The scientific case was overwhelming. Such an array could see into the hearts of star-forming regions, with the same resolution as the HST but unhindered by dust obscuration. It could detect the most distant star-forming galaxies and follow their evolution over the history of the universe. And it could study their detailed chemistry using a forest of molecular spectral lines. Operating at millimeter and submillimeter wavelengths, it would be at the crossroads of radio and optical/infrared astronomy, so it attracted great interest from both communities. The technologies required to build high-precision mobile antennas that could operate in the open at an altitude of 5,000 m, using receivers with sensitivities approaching the quantum limit, were rapidly being developed.

Happily, similar developments were underway at the same time in both the United States and Japan. There was close communication between the three groups, and, given the magnitude and cost of the projects, collaboration and finally a merger of all three took place. The complexities of organizing and running a single huge observatory were ultimately overcome, resulting in the magnificent ALMA observatory that was inaugurated in 2013.

Peter Shaver
European Southern Observatory (retired)
Woollahra, Australia

Japan's Large Millimeter/Submillimeter Array

Another project that paralleled those of the MMA and LSA was being developed in Japan. At first, it was called the Large Millimeter Array (LMA). Experience with the NRO millimeter interferometer, in the Japanese mountains a few hours from Tokyo, had made it clear that a much larger collecting area and longer baselines were required to study the astronomical topics of current interest. As early as 1983, the Japanese had proposed to increase the number of 10 m diameter antennas in their array from 5 to 30.

One evening during the International Astronomical Union (IAU) Symposium #115 titled *Star Forming Regions*, which was held in Tokyo in 1985, Vanden Bout broached the topic of possible cooperation between the MMA and LMA with Japanese colleagues Masaki Morimoto and Norio Kaifu. Morimoto was a pioneer of radio astronomy in Japan and Kaifu a rising star in Japanese astronomy. Although interested, the Japanese radio astronomers said it was too early for any formal arrangement. Their focus was on the Subaru (optical/infrared) Telescope, which was planned to be constructed in Hawaii in the early 1990s. The year after the IAU Symposium, Vanden Bout visited the NRO headquarters to continue the discussion with Morimoto and his colleague Masato Ishiguro on partnering with the MMA, and to see their facility first-hand.

By 1987, the Japanese had a plan[14] that called for (50) 10 m antennas to be built on the best site that could be found in the world. The project name was changed to the Large Millimeter/Submillimeter Array (LMSA). The emphasis now was on operating at submillimeter wavelengths, which the Nobeyama site could not support. Consequently, the search for the LMSA's potential site began in 1992, both in Chile and in Hawaii. Having the LMSA sited near the Subaru Telescope, on Maunakea, was attractive due to the potential cost savings associated with shared facilities; each of the NRAO and ESO projects had also been through this thought process in their respective searches. Furthermore, travel from Japan to Hawaii is much more convenient than travel to Chile, which is normally accomplished by going via the United States or even Europe. It took many years before the choice was made, but over those years the search increasingly focused on a site in northern Chile inland of Antofagasta, called Río Frío, at an elevation of 4,200 m, in the Domeyko mountain range roughly halfway between Paranal and the Andes Mountains inland. They also studied a site called Pampa la Bola (the literal translation is *Boron Prairies* as this was the prime location for boron deposits in Chile) where they would later place a single-dish submillimeter telescope. A potential site in northwest China was searched, but they could not find a better site than the one in Chile.

During those years, there were several meetings between NRAO and NAOJ, relevant to their respective MMA projects. At one such meeting, in the New Otani Hotel on Kimana Beach in Honolulu, Don Hall, director of the IfA, appeared unannounced. He was eager to have the IfA enjoy special access to a major MMA on Maunakea and was annoyed that the two observatories had arranged a meeting in Hawaii to discuss their plans without giving him notice. The oversight was deliberate – the meeting was held to discuss siting the projects in Chile, not on Maunakea. After Hall was told of NRAO's intention to site the MMA in Chile, he said he would not try to block the decision. However, were NRAO to take the LMSA with them to Chile, he would have Hawaii's Senator Daniel Inoue stop funding for the MMA in the US Congress. Later, Japan did choose Chile for the LMSA site, but if Hall ever called Inoue, the senator did not act.

The IAU Colloquium #140 – *Astronomy with Millimeter and Submillimeter Wave Interferometry* held at Hakone, Japan in 1992 – was the first international meeting devoted to millimeter and submillimeter wave interferometry, and provided a good opportunity to develop ideas about collaboration among large interferometer projects. On 8–9 March 1994, Vanden Bout and Brown visited NAOJ in Mitaka, and collaboration between the MMA and LMSA was officially discussed for the first time. An agreement[15] between NRAO and NAOJ was signed on 12 June 1995 to collaborate in the study of potential sites in Chile for the LMSA and MMA. On 8–10 November 1995, Vanden Bout participated as a member of the NRO visiting committee. The committee endorsed the LMSA project with enthusiasm.

Just after the meeting, Bob Brown and Peter Napier visited the NRO with Vanden Bout to see, among other areas, the electronics laboratories. The photograph in Figure 5.1 was taken at the Kiyosato Resort just prior to this visit. They were struck by the contrast between NRAO, where antenna maintenance and electronics development were done by observatory staff, and NRO, where these functions were accomplished by employees of Japanese industry, under contract

Figure 5.1 Left to right: Paul Vanden Bout, Peter Napier, Masato Ishiguro, and Bob Brown. Courtesy of M. Ishiguro, reproduced by permission.

from NRO and supervised by NRO scientists. Soon after, a workshop was held on 16–19 March 1997 in Tokyo called *Millimeter and Submillimeter Astronomy at 10 MilliArcSeconds Resolution*, under the cooperative research program supported by the Japan Society for the Promotion of Science and NSF. The workshop was well attended with participants from Europe as well as the United States and Japan. Discussion at the workshop was based on the feasibility of a conceptual array – the Atacama Array – that consisted of (40) 8 m diameter MMA antennas and (50) 10 m diameter LMSA antennas, the latter capable of observing submillimeter wavelengths. The arrays were either to operate independently or in a combined mode, a model that was later adopted when the Japanese entered ALMA. The sense at the conclusion of the workshop was that the United States and Japan might well have the basis for a partnership.

The Shimogamo Saryo sits on the bank of the Takano River in the heart of the Kyoto geisha district. Founded in 1856 as an *ochaya*, or traditional tea house, it continues today as a Michelin Star restaurant celebrated for its traditional Kyoto cuisine. It was the scene for a dinner in August 1997 that celebrated the developing partnership between NAOJ and NRAO. Keiichi Kodaira, Norio Kaifu, Paul Vanden Bout, Bob Brown, Bob Dickman, and their wives enjoyed an elaborate meal and copious amounts of sake. The dinner had been arranged by Masato Ishiguro, who did not attend. He secured the reservation and negotiated a reduced price by arguing that the guests were distinguished astronomers and the conversation that night would lead to something important for the national astronomy program of Japan (the Atacama Array).

On 24 March 1998, an agreement[16] to cooperate in the protection of the Chajnantor and nearby Pampa la Bola sites, technology development, and discussion with ESO to include the LSA in a three-way partnership was signed at a meeting in Hawaii.

The MMA and LSA Become One Project

Around the time of the Tokyo workshop, the thinking in Europe, and specifically at ESO, was coming to the conclusion that an international partnership that included countries beyond Europe would be preferable when building a large millimeter array. Giacconi had grown to appreciate the scientific potential of radio astronomy, in particular, millimeter/submillimeter astronomy. The endorsement by the STC and the strong justification for a large millimeter array given by Woltjer were all Giacconi needed as reason to proceed. He was aware of the discussions NRAO had with the Japanese about a merger of the MMA and LMSA they called the Atacama Array, and he did not want to be left out. Furthermore, he needed a means of preserving the current level of the ESO budget, which was poised to decline with the completion of the VLT, until the time OWL[17] could be started. The LSA would also preserve ESO's expert workforce, which had been built up to complete the VLT. In April 1997, the report[18] of the IRAM/ESO/OSO/NFRA group became available. It reiterated the scientific case and set out the specifications for the LSA. At the 4–5 June 1997 meeting of the ESO Council,[19] the LSA was discussed, and Giacconi was encouraged to "*continue exploratory talks ... on an international level.*" In early June 1997, he called Vanden Bout and proposed a meeting to discuss merging the MMA and LSA.

The call took Vanden Bout by surprise. He had failed to keep up with the changing attitude toward millimeter astronomy at ESO. His last conversation with Giacconi had been the previous October, at a meeting of the Spanish Astronomical Society in San Sebastián. After his talk on the MMA, Giacconi had chided him for trying to pick off European countries one at a time as partners of the MMA. There was no hint that he was warming to the idea of merging the LSA with the MMA. In retrospect, a merger of the European and US projects, not to mention the Japanese project, was inevitable. It would have been crazy to build three independent arrays, all in Chile, all with the same scientific goals. But at the time, the only partnership being discussed was the Atacama Array between the United States and Japan.

Now Europe wanted to talk. It was important to act promptly, as NRAO had learned that the NSF was seriously considering funding the design and development work for the MMA. If the project were to change significantly in scope, NSF needed to be informed. There was a need to meet as soon as possible. Giacconi felt rushed when told of the urgency and accused Vanden Bout of pulling the "train is leaving the station" ploy. Significantly, both agreed that the merged project would be the sum of the MMA and LSA, not one of those projects at half the cost to each side. Their conversation concluded with two remarks that came to define the management structure of ALMA. Giacconi said, *"We will not be less than 50 percent partners." "Neither will we,"* was the reply. Giacconi agreed to come to NRAO as soon as it could be arranged.

The meeting took place within the month, on 25–26 June 1997 in the NRAO Charlottesville auditorium. The Europeans and Americans faced each other across a large U-shaped arrangement of tables, the two sides opposite one another.[20] Neal Evans, from U. Texas (Austin), thought it looked like the negotiation of a truce between two Mafia dons. But the meeting was cordial, and agreements were easily reached on the top issues. By the morning coffee break of the second day, it had been decided to study a nominal array of sixty-four 12 m diameter antennas, capable of submillimeter observing, on the high Chajnantor site near San Pedro de Atacama. A journal[21] kept by Bob Brown documented in handwritten notes the comments by various participants. There were no serious arguments because everyone got what they wanted: the Europeans a large collecting area, the Americans the broadband operating capability that the high site supported. Without any authority to do so, Giacconi and Vanden Bout signed the resolution[22] shown in Figure 5.2 to cooperate in the exploration of a merger of the LSA and MMA. Ed Churchwell is quoted[23] as saying, *"Although there are outstanding unresolved questions, I believe this was a watershed meeting."* Torben Andersen[24] recalls riding to the airport with Giacconi after the meeting: *"... he was quite happy with the outcome of the meeting, because it was finally cemented that ESO would be the counterpart in Europe."* The US side was also happy. The MMA was now part of an international project as NSF required.

The agreement toward this merger of the MMA and LSA had come on the heels of the very successful workshop in Tokyo which had led to the Atacama Array concept, an idea shared by NRAO and NRO. The Japanese, understandably, were shocked and felt blindsided by the signed resolution between NRAO and ESO, with its mere mention of continued cooperation with the Japanese project. To make matters worse, Ishiguro was visiting the NRAO electronics laboratory while the resolution was being hashed out nearby. It was an acutely embarrassing situation, caused by Vanden Bout's failure to give Ishiguro advance notice of the meeting. He met with Ishiguro to apologize and to assure

RESOLUTION

Whereas the development of millimeter-wavelength astronomy has shown the potential of large millimeter interferometric arrays for revealing the origin and evolution of stars and planetary systems, of galaxies, and of the Universe itself; the communities in the United States and Europe have proposed the construction of the Millimeter Array (MMA) and the Large Southern Array (LSA), respectively; and there is an opportunity through cooperation to achieve more than either community planned; we, as the observatories responsible for these projects and with the support of our communities, resolve to organize a partnership that will explore the union of the LSA and MMA into a single, common project to be located in Chile. Specifically, this partnership will study the technical, logistical, and operational aspects of a joint project. Of particular importance, the two antenna concepts currently under consideration will be studied to identify the best antenna size and design or combination of sizes to address the scientific goals of the two research communities. In doing so we will work through our observatories, utilizing the expertise in millimeter astronomy resident in research groups and institutions in our communities. Finally, we recognize that there are similar goals for millimeter astronomy in Japan, and cooperative activities with that project will continue.

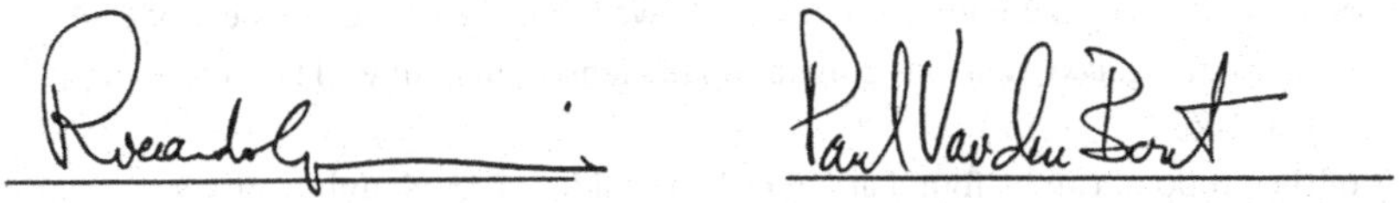

R. Giacconi   P. Vanden Bout

European Southern Observatory   National Radio Astronomy Observatory

26 June 1997

Figure 5.2 The resolution between NRAO and ESO in which the parties agreed to explore a merger of the LSA and MMA into a common project. Credit: NRAO/AUI/ NSF, CC BY 3.0.

him of NRAO's and ESO's sincere commitment to bring Japan into the project. However, it would be seven years before Japan officially joined the partnership. Finally, on 14 September 2004, Japan formally entered the ALMA partnership under a provision of the Bilateral ALMA Agreement, bringing to ALMA a compact array of twelve 7 m antennas, four 12 m antennas, a correlator, and three additional receiver bands for the large array. Later, in 2015, the Trilateral ALMA Agreement was signed, formally committing NSF, ESO, and Japan, to a three-way partnership for the operation of ALMA.[25] Two books[26] have been written about ALMA from a Japanese perspective.

Needless to say, news of the resolution with ESO was well received at the NSF. Never mind that it was unauthorized and unofficial. The MMA now satisfied the requirement to be an international project. In May of the following year, the National Science Board authorized $26 million from the Major Research Equipment account for MMA Design and Development for FY1998–2000 upon *"formal establishment of an international partnership."* It would be two more years before the NSF signed an MOU with ESO that officially merged the LSA and MMA. And another two years before the first construction money was made available. Still, several hurdles had been cleared.

Notes

1 Brown and Vanden Bout (1991) reported data that suffered from systematic noise due to the poor observing conditions at Kitt Peak that (and every) summer. They were forced to observe through the fabric cover of the telescope dome some of the time. The result was an overestimate of the CO line strength. (Later observations by Radford et al. (1996) using the 12 Meter Telescope gave the correct line strength, in agreement with measurements at several other millimeter telescopes.) That and a faulty analysis of the data led them to conclude that they had detected a supermassive galaxy in formation. Further observations and a proper analysis by Solomon et al. (1992) showed that the data could be modeled by a large star-burst galaxy that was gravitationally lensed, magnified 11 times by an intervening galaxy.

2 A copy of the proposal can be found at NAA-NRAO, MMA, MMA Planning, Box 3.

3 The referee's reports can be found at NAA-PVB, ALMA, ALMA: The Story of a Science Mega-Project. https://science.nrao.edu/about/publications/alma.

4 The letters can be found at NAA-PVB, ALMA, ALMA: The Story of a Science Mega-Project. https://science.nrao.edu/about/publications/alma.

5 Brown's notes taken at the SEST Users Meeting can be found at: https://library.nrao.edu/public/memos/alma/misc/ALMAU_10.pdf.

6 The membership of the working group was Roy Booth (OSO), chair, Lars Bååth, (OSO), Peter Dewdney, (DRAO), Dennis Downes (IRAM), Michael Grewing (IRAM), Stepháne Guilloteau (IRAM), Frank Israel (Leiden U.), Peter Shaver (ESO), Tom Wilson (MPIfR), and Masato Ishiguro (NRO). Ishiguro was unable to attend this first meeting. From Booth (1994).

7 Downes presented his vision of the future of millimeter astronomy at the XVII ESLAB Symposium, held at ESTEC in Noordwijk, the Netherlands, 10–14 May 1993 (Downes, 1994), p. 133. He argued that the future lay in the millimeter band and that a mm array capable of studying galaxies in the early Universe should have a collecting area of 10,000 m^2, an angular resolution of 0.1 arcseconds at 2.6 mm wavelength, and a site with 10 × 10 km area of flat terrain above 2,000 m elevation in a dry climate.

8 NAA-PVB, ALMA, ALMA: The Story of a Science Mega-Project. https://science.nrao.edu/about/publications/alma.

9 The report with appendices by P. Shaver to the STC at their meeting of 30–31 October 1996 can be found at: NAA-PVB, ALMA, ALMA: The Story of a Science Mega-Project.

10 Ibid., Appendix D.

11 Ibid., Appendix F.

12 Beckwith to Vanden Bout, private communication.

13 Shaver to Vanden Bout, private communication.

14 A description of the LMA was given in 1992 at IAU Colloquium 140 (Ishiguro, M., 1994). Instead of adding 25 antennas to the NMA, the concept was "*extended to an array that would realize sub-arcsecond resolution imaging at very high frequencies. The LMA may consist of 50 10-m antennas and will be covering observing frequencies from 35 to 500 GHz (possibly 650 and 800 GHz).*" Site testing for the LMA was focused on northern Chile, as atmospheric data on Maunakea was available from other groups. By 1997, the project, now the LMSA, was presented at IAU Symposium 170; see Ishiguro (1997). See also Ishiguro (1998).

15 The agreement to jointly study potential sites can be found at NAA-NRAO, ALMA, ALMA Multi-Institutional Agreements. https://science.nrao.edu/about/publications/alma.

16 The agreement between NRAO and NAOJ to cooperate in the study and protection of the Chile sites can be found at NAA-NRAO, ALMA, ALMA Multi-Institutional Agreements.

17 Beginning with Catherine Cesarsky's tenure as ESO Director General, OWL evolved into the Extremely Large Telescope (ELT), which has a 39.3 m diameter aperture and is at the time of the writing of this book under construction.

18 The IRAM/ESO/OSO/NFRA report can be found at: NAA-NRAO, MMA, MMA Planning, Box 6. https://science.nrao.edu/about/publications/alma.

19 Claus Madsen notes this in his history of ESO (Madsen, 2012), p. 376.

20 There is no complete record of who attended the meeting, but from Brown's journal notes and from the memories of those who confirmed to the authors that they had attended, we can state that the following were present for the United States: Paul Vanden Bout, Bob Brown, Peter Napier, Neal Evans, and Ed Churchwell. Unfortunately, the only available records of the meeting yield an incomplete list for the US side. Madsen (2012) lists those present from Europe as: Riccardo Giacconi, Peter Shaver, Roy Booth, Stepháne Guilloteau, Torben Andersen, Dietmar Plathner, and François Viallefond.

21 Brown's journal can be found at: NAA-RLB, Calendars and Journals.

22 A copy of the resolution is in NAA-NRAO, MMA, MMA Planning, Box 6. https://science.nrao.edu/about/publications/alma.

23 Madsen (2012), p. 376.

24 T. Andersen to P. Vanden Bout, private communication.

25 A condensed account of the development of the MMA, LSA, and LMSA, and their eventual merger into ALMA was given at the *Dusty Universe* conference (Vanden Bout, 2005).

26 For accounts of the ALMA project from a Japanese perspective see Ishiguro (2009) and Yamane (2017).

6

Organizing ALMA

"Organizing is what you do before you do something,
so that when you do it, it's not all mixed up."

Unknown – misattributed to A.A. Milne

The agreement of 25–26 June 1997 between ESO and NRAO stated a *"resolve to organize a partnership that will explore the union of the LSA and MMA into a single, common project …"*. The organization proceeded in stages, with the establishment of working groups, advisory committees, and eventually, the appointment of a board to oversee the project. The organizational tasks were different in Europe and the United States for their respective projects. Until the assignment of tasks to Integrated Product Teams (IPTs) by the ALMA Coordinating Committee (ACC) in 2000, tasks were addressed on each side by separate working groups. NRAO was about to receive MMA design and development funding from the NSF, whereas the ESO Council had not yet received a proposal to build the LSA. Despite the lack of a formal, coordinated management structure at the beginning, the groups worked well together in an atmosphere that was extraordinarily collegial. The recommendations of the working groups ultimately came to define ALMA. However, at the same time that the working groups were discussing science requirements and array specifications, a completely different challenge for the budding partnership became urgent, namely a proposal by a commercial firm to build a gas pipeline directly across the Chajnantor site.

Gaz Atacama

A company called Gaz Atacama (GA) had a plan to lay a pipeline from Argentina to Tocopilla, just north of Antofagasta on the Chilean coast. The pipeline would supply natural gas from Argentina to plants generating electricity for the Chilean mining industry. NRAO became aware of the plan when GA requested a right of way from the Ministry of National Assets, "Bienes Nacionales" (BN), the Chilean government agency that owned the land. As discussed in Chapter 4, AUI had already secured the mining claim in 1996 and CONICYT had also requested a very large land concession for the MMA on behalf of NRAO/AUI. Paulina Saball, the Undersecretary of BN asked both interested parties – NRAO and GA – to reach an agreement or she would pick a winner. The negotiations began in November 1997. The NRAO team was led by Eduardo Hardy and included Peter Napier, the MMA project manager, and Jeff Kingsley. The parties agreed to separate their projects by a distance of 200 m, that is, GA had a buffer zone 400 m wide. The buffer zone would be violated in three places where the antennas would be transported across the pipeline. At those spots, provisions were to be made to reinforce the pipeline. It was agreed that the pipeline monitoring equipment, which was to transmit data by radio, would avoid doing so in the area of the site. Finally, a tap in the pipeline was promised by GA so that the MMA could draw gas to generate electricity. A legally binding agreement[1] was signed in April 1998. But the rights of the two parties to the land would not become legal until 2004, following long negotiations with the Chilean government.

Although ALMA never used any gas from their pipeline tap, construction of the pipeline did provide a bonus. In the late 1970s and 1980s, the Chilean Army had laid mines along Chile's long border with Argentina, as well as with Bolivia and Peru to the north, fearing possible invasions during times of conflict and chaos. Chile ratified the Ottawa Convention (the Anti-Personnel Mine Ban Treaty) in 2001 and the government set to work removing the mines they had previously buried. It was a long process because in the intervening years, many of these mines had moved with soil that washed down the mountainside gullies and the mine locations were no longer accurately known. As a precaution, GA conducted a sweep for mines along the pipeline route and the project could proceed with confidence that the area was safe. One mine was found by accident when one of the GA bulldozers drove over it in a ravine near the Jama Pass highway far below the site. The driver lost an eardrum in the explosion. The pipeline project had an unexpected outcome. Before gas began flowing, the Argentine government decided not to export their gas after all.[2]

Land Concession

Concluding an agreement to build and operate the MMA, and then ALMA, on the Chajnantor site would prove to be a long and difficult process.

Essentially no progress was made[3] until the Ministry of Foreign Affairs took eventual control on 26 July 2001. In a large meeting that day, chaired by Ambassador Luis Winter of the Ministry of Foreign Affairs, no less than 14 different entities were told to come to an agreement with the ALMA representatives on the conditions for their approval. The U. Chile had already concluded agreements with AUI, first for the MMA and later for ALMA, granting access to the land; it was indicated that site studies, construction, and operation of an observatory were permitted in exchange for the 10 percent of observing time in compliance with Law 15172, that was to be administered by U. Chile for the benefit of all Chilean astronomers. Such an agreement would be congruent with the agreement between ESO and Chile, signed in 1996, that gave 10 percent of the observing time on ESO's telescopes to Chilean astronomers.[4] There was little further progress until 19 February 2002 when the Undersecretary of the Ministry of Foreign Affairs, Cristián Barros, asked both sides for the list of requirements essential to ALMA and to the agencies involved. On 20 May 2002, negotiations under Barros began. Vanden Bout was the head of the negotiating team, but Daniel Hofstadt for ESO and Eduardo Hardy for AUI did all the real work.

Following the successful negotiations, the Treasury Department issued two decrees, first for the MMA, specifying AUI's exemptions from import duties and value-added taxes, among other details; the second for ALMA, specifically redefining the MMA as the *fraction of ALMA belonging to AUI in Chile.* The Labor Department agreed to follow inspection rules consistent with immunities granted by the Ministry of Foreign Affairs. A land concession to Radioastronomía Chajnantor, Limitada (RCL), a joint company held by ESO and AUI, was granted by BN in return for annual payments to two funds, one for Region II and the other for CONICYT. BN also sold the land for the Operations Support Facility (OSF), more than 2,000 m lower in elevation and 28 km away from the chosen array site, to RCL. The Ministry of the Environment approved the environmental impact study, which contained 91 commitments to be kept during construction and operation of ALMA; those commitments involved a number of other agencies and entities in Chile. It was also important to get the approval of the San Pedro de Atacama mayor, Sandra Berna. The governor of Chile's Region II, Jorge Molina Cárcamo gave his approval after the agreement was reached on annual payments by ALMA to a social development fund for the San Pedro region. CONICYT would administer annual payments to a fund for the development of astronomy in Chile. The Forestry Department authorized a right of way from the OSF to Route 23, the primary access, in exchange for a one-time payment. The

Figure 6.1 Left panel: President Ricardo Lagos greeting the local residents of the San Pedro community. Right panel: President Lagos giving his speech on astronomy. One of the attendees appears to have heard enough. Credit: Eduardo Hardy; NRAO/AUI/NSF, CC BY 3.0.

Agricultural Service gave its approval after ESO/AUI made a commitment to protect the flora and fauna. The Department of National Monuments granted its approval after ESO/AUI promised to protect historical sites. The agreement with GA that was discussed earlier was also one of the requirements for gaining legal access to build and operate ALMA. The signing of decrees was celebrated at two receptions. The first was on 25 July 2003, in San Pedro de Atacama. President Ricardo Lagos can be seen in Figure 6.1 greeting the assembly before he gave a speech on the wonders of astronomy. Eduardo Hardy also gave a speech in which he thanked President Lagos and Jaime Ravinet, the head of BN. Ravinet then signed the degree granting the right to build and operate ALMA, followed by Daniel Hofstadt and Eduardo Hardy for ESO and NRAO/AUI, respectively.

The second reception was held on 21 October 2003 at the Ministry of Foreign Affairs. An agreement allowing ESO to build and operate ALMA on the Chajnantor site was signed by Maria Soledad Alvear, Minister of Foreign Affairs, and Catherine Cesarsky, who had followed Giacconi as ESO Director General. The agreement added ALMA to ESO's existing Convenio with Chile for its optical telescopes. A photograph recording the signing of the agreement is shown in Figure 6.2.

On the Chilean side, several individuals stood out in the effort to make the land negotiation process successful, and deserve mention. Leo Bronfman and his colleagues at U. Chile were notably helpful, especially at the start when NRAO first came to Chile looking for an MMA site. Hernán Quintana, who

Figure 6.2 Left to right: (unidentified), Catherine Cesarsky, ESO Director General, Maria Soledad Alvear, Foreign Minister of the Republic of Chile, (unidentified). Credit: ALMA/ESO/AUI, CC BY 4.0.

first pointed to the Chajnantor area as a possible MMA site, later became anxious that the agreement with the U. Chile would limit access to MMA time for him and his colleagues at U. Católica. His concern reinforced the position of the ALMA partners that CONICYT would have the responsibility for administering ALMA observing time on behalf of *all* Chilean astronomers. As already noted, Cristián Barros, Undersecretary of the Ministry of Foreign Affairs, was a key figure in the early negotiations. He took over when things appeared to be bogged down, demanding that all the parties come to an agreement, and putting Ambassador Luis Winter, his right-hand man, in charge. By a stroke of good fortune one of the authors (Dickman) was at the time a US Embassy Fellow at the American Embassy in Santiago, on temporary leave from NSF. At his request, the US Ambassador, William Brownfield, convinced the Governor of Region II that ALMA was not flush with money, like a mining company, but a scientific enterprise that could not afford the large payments that had originally been requested. The efforts of these people in particular, as well as the goodwill of all involved, led to a successful conclusion, however long it took to get there. The process leading to these agreements is shown as a flow chart in Figure 6.3 and a map of the land in question is shown in Figure 6.4.

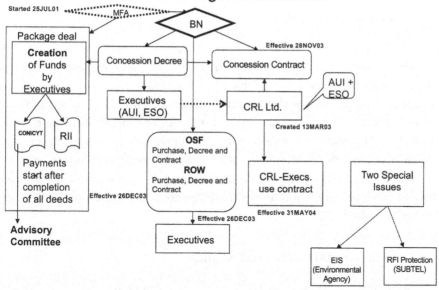

Figure 6.3 A flow chart showing the steps taken in acquiring the ALMA site. The label "CRL" should be "RCL" for Radioastronomía Chajnantor, Limitada. Credit: Eduardo Hardy, NRAO/AUI/NSF, CC BY 3.0.

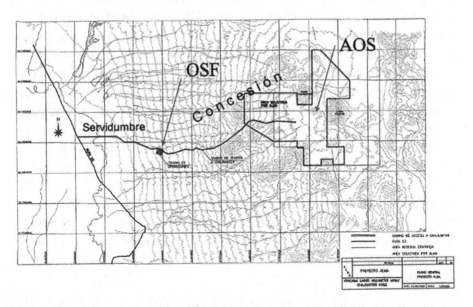

Figure 6.4 The land concession sought for the MMA and then ALMA, showing the large 17,000 hectare Array Operations Site (AOS), the right-of-way (labeled "Servidumbre") to the Operations Support Facility (OSF), and then to Highway 23 running from San Pedro de Atacama to the village of Toconao. The OSF occupies a 1 km × 1 km piece of land that was purchased outright. Elevation contours and runoff ravines are shown in gray. Credit: Adapted from a topographic map of the Instituto Geográfico Militar; NRAO/AUI/NSF, CC BY 3.0.

Early Joint Activities

There were joint technical, science, and management working groups that began as early as 1997, but they were only "joint" in the sense that groups on both sides of the Atlantic kept each other informed of progress and, on occasion, attended each other's meetings. The only published reports of these working groups are those that appear in the LSA Feasibility Study, and those are of the European side. Bob Brown, Paul Vanden Bout, Peter Shaver, and Michael Grewing met in Munich to discuss management options, but they did not issue a report either. The highlight of their meeting seems to have been an evening concert featuring Daniel Barenboim conducting the orchestra while playing the piano.

Defining ALMA in the United States

A Revised Design and Development Plan – The long delay before design and development funds were finally (and only verbally) promised by NSF Director Neal Lane meant an updated plan was needed. By early 1997, too much had already changed. There also was a sense of urgency, now that NRAO and ESO had agreed to explore a merger of their projects. That updated plan[5] was submitted in January 1998, as the *MMA Program Plan – Design and Development, Volume 2*. It was similar to the first plan in structure, with updated discussions of the plan's elements. A very significant change revealed in this updated plan was the switch from forty 8 m antennas to thirty-six 10 m antennas. This array was regarded as the one that would meet the many enumerated scientific requirements, while also ensuring a scope that the United States could manage alone should the proposed merger fail. It came to be called the *US Reference Design for the MMA*. Another significant change was to specify the two sites that were then under consideration: Maunakea in Hawaii, and the Atacama Desert in Chile. No mention was made of the former three US continental sites. The revised plan included extensive material as attachments that documented the results of technical studies. The more detailed work breakdown structure (WBS) led to a budget of $26M for a design and development program, to be spread over four years of construction.

The Reference Design for the MMA – Increasing the collecting area of the MMA by 40 percent while only cutting the number of antennas by 10 percent had increased the construction cost well beyond the $120 million total in the MMA proposal. There was also inflation over the intervening years to take into account. Dickman, with strong support from Bob Eisenstein, the NSF Assistant Director for MPS, asked for a cost estimate of the expanded array.

The documents were prepared and submitted to the NSF as requested. Not unexpectedly, the cost had escalated significantly. Taken aback by the magnitude of the increase, the NSF scheduled a so-called Lehman Review, named after Daniel Lehman, who had conducted in-depth reviews of science projects for the Department of Energy. Lehman Reviews were considered the gold standard for audits of large project feasibility and cost. This was a life and death moment for the MMA. Failing the review would possibly mean the end of the project, whereas a strong endorsement would mean renewed confidence at NSF for pushing forward.

The review took place 8–9 July 1999. The panel was chaired by John Peoples, who had just stepped down as director of Fermilab to take charge of the Sloan Digital Sky Survey. It included other experts in large project management, as well as leading scientists and engineers in radio astronomy. Sessions were attended by NSF officials, NRAO staff, and observers from other projects around the world. One attendee was AUI trustee Paul Gilbert, a leading expert on the construction of tunnels at the engineering firm Parsons Brinckerhoff, and a veteran of the Super Conducting Super Collider debacle. He kept a close eye on the MMA for AUI and gave sound advice to the project. Ultimately, a very complete report[6] was issued. The panel examined the cost estimates in detail, found them to be reasonable, and concluded that the US Reference Design for the MMA could be built for around $400 million. Beyond this overall finding, the panel made numerous recommendations for improvements in the management structure, in the WBS, and for actions in the areas of site development, electronics, and computing. The confidence of the panel in this cost estimate and the helpful criticism in the report provided exactly what was needed to advance the MMA within the NSF. The US Reference Design for the MMA evolved into the US contribution to ALMA. Backed by the Lehman Panel's report, the cost of the MMA was reset in NSF's thinking.

Further Steps in the Merger

Studies proceeded amicably but independently by the respective groups in both Europe and the United States in order to explore considerations to be addressed in a merger of the LSA and MMA. The US effort was led by a small cadre of NRAO staff members. The standing MMA advisory committees were composed of university-based astronomers and engineers. Receipt of funding for design and development would allow for the formation of a larger team at NRAO, as well as financing for support tasks by the university members of the Joint Development Group (JDG). The JDG was set up by NRAO to involve university millimeter astronomy groups, principally at the Owens Valley Radio Observatory and the participants in the Berkeley-Illinois-Maryland Association, in the design of the MMA.

In Europe, on 17–18 December 1998 an MOU[7] was signed between ESO and the funding agencies in several of its member countries wishing to participate in LSA design and development to form the European Coordinating Committee (ECC). Initially, the interested countries forming the ECC were France, Germany, The Netherlands, and the United Kingdom. Later, Sweden and Spain joined. Each of the signatories pledged support in money and labor, with the goal to agree on specifications of an array that merged the MMA and LSA. There was a push toward matching the effort to be funded by NSF in the United States. The MOU also established the European Executive Committee (EEC) to manage the activities of the MOU, and the European Negotiating Team (ENT) to negotiate an MOU with the NSF for the design and development phase of the joint project. One of the first actions of the ECC was the appointment of Dick Kurz and Stepháne Guilloteau as European Project Manager and Project Scientist, respectively.

The developing partnership between Europe and the United States to build a merged array needed a joint management structure. In April 1999, an MOU[8] specifying all the design and development tasks required for an array of sixty-four 12 m antennas on the Chajnantor site was concluded between NSF and the principal parties to the ECC. It was signed by all the parties in June 1999. The MOU covered Phase 1 of the merged project, with the objective of completely defining the work to be carried out in Phase 2, including the definition of all the scientific and technical requirements and the management approaches, as well as a WBS, schedule, and costs. The MOU for Phase 2, which addressed actual construction, was to be negotiated and signed by 31 December 2000.

The Phase 1 MOU contained contractual language that was necessary, if tedious to read. However, along with the legalese there was an item of great significance, namely the appointment of the ACC, whose charge was to supervise all the activities under the MOU. Later, it would provide a model for the ALMA Board. The ACC was expected to meet at least twice per year, although in practice it met as often as monthly. It was to elect its own chair, with that position alternating between the United States and Europe, and had an initial membership[9] of 12, with six representatives from North America (NA) and six from ESO.

The work during Phase 1 was supervised for NA by the NRAO MMA Project Director, Bob Brown, and for ESO by ECC Project Manager, Dick Kurz. The Project Scientists were Al Wootten (NRAO) for the United States and Stepháne Guilloteau (CNRS) for Europe. This group together with the NRAO Director and ESO Director General formed the ALMA Executive Committee (AEC).

Defining ALMA in Europe

LSA/MMA Feasibility Study – The first formal step in Europe to define ALMA was a study of whether it was even feasible to combine the MMA and LSA. The feasibility study involved nine working groups, and was run by the LSA Board,[10] which had led the earlier development of the LSA. The working groups were largely composed of European scientists and engineers, although the antenna working group added Peter Napier of NRAO as an unofficial member to present the 10 m and 12 m antenna options that would be considered for the MMA. The report[11] entitled *LSA/MMA Feasibility Study* concluded that a combination of the MMA and LSA would indeed be a powerful facility for addressing ESO's larger science goals, and could be built within a reasonable budget. It formed the basis for the proposal that was later prepared for the ESO Council.

ESO ALMA Proposal – This massive document of 676 pages in six volumes is the most complete and detailed description of ALMA on record.[12] Dated December 2000, it was drafted at a sufficiently late date to have the benefit of over three years of planning. The science case for ALMA is superbly presented. Its six volumes represented the defining statement of the whole project at the time that it was submitted to the ESO Council for approval.[13]

Dividing the Effort

At the suggestion of Riccardo Giacconi, the ACC agreed that contributions to Phase 2 would be in the form of completed work packages, with the amounts of actual cash contributions kept to a minimum. That is, the project would be divided into deliverables: antennas, site development, receivers, signal correlator, etc. with each deliverable assigned an estimated value. The sum of the values was to be the same for Europe and NA. Moreover, once a party had agreed to provide a deliverable, they were obligated to do so, no matter what the final cost came to, although it was obvious that if things got out of hand there would be a renegotiation of value. This process required the construction of a detailed WBS. To get started, Brown and Kurz made rough estimates of the value of the major packages and, knowing the preferences of the respective partners, divided them as best they could to assure equity. The largest package was the antennas. Both sides wanted to provide the antennas, but it was unreasonable and unrealistic to expect that one partner would, for example, provide all the antennas and the other provide most of the rest of the array components. So, that work package was split, with each side agreeing to contribute half of the antennas. When it came to assigning the site construction, ESO argued that its long experience in Chile was an obvious reason for them to do all the site work. But NRAO was loath to have ESO be the single face of the project on the Chilean site. Again, the two sides agreed to split the package, with NRAO taking responsibility for the high site work at 5,000 m and ESO taking on the mid-level OSF. Negotiations about the other deliverables followed similar lines. For example, receiver inserts were divided between Europe and the United States, their cryogenic containers being built in England but the integration of the receiver inserts into the container carried out at NRAO. The final division of overall effort was hammered out in a retreat at Abingdon, England, in the spring of 2000. The chief participants were Massimo Tarenghi, Bob Brown, Dick Kurz, and Masato Ishiguro. It was ratified in the MOU for Phase 2.

It should be noted that the Japanese participated fully in these discussions even though Japan had not yet joined ALMA. The final WBS concluded at the Abingdon retreat, attended by Masato Ishiguro and Tetsuo Hasegawa, included all the contributions to ALMA expected from Japan, except for an additional compact array of 7 m antennas, an item left for further negotiation. Their participation made the final negotiation of the terms for their entrance into ALMA much easier than it would have been had they been excluded. It was another example of the collegial relations that characterized the global millimeter astronomy community at that time.

ALMA Coordinating Committee

The ACC met 25 times from March 1999 to December 2002. Thirteen of those meetings were in-person at locations around the world (London, Munich, Paris, Santiago, Tokyo, Venice, and Washington DC). Twelve were telecons, with audio/video technology that preceded Zoom® by many years and could prove frustrating to the participants. Besides the routine of reports, scheduling the next meeting, and approving the minutes of the last, there were key issues of concern to the ACC over which much effort was expended. These included: (1) monitoring the division of effort; (2) advising on the negotiations for the site, with special attention to the differing institutional requirements for ESO and AUI; (3) deciding who would employ the local workforce; (4) selecting a location for the ALMA headquarters in Chile; (5) negotiating the entrance of Japan into ALMA; and most importantly, (6) managing the negotiation of the ALMA Agreement between ESO and NSF. A full discussion of all the meeting agendas, discussions, conclusions, and minutes is beyond the scope of this book. In the following text, we give brief summaries of the more significant meetings.

30 March 1999, Garching – This meeting at ESO Headquarters was actually a meeting between NSF representatives and the ENT rather than an official ACC meeting. In fact, the attendees became the ACC at its first meeting the following June. A significant step at this meeting was the adoption of the name ALMA for Atacama Large Millimeter Array. Bob Brown reported the results of an informal poll in which he had requested votes for suggested names. ALMA came out on top, a joint suggestion from Paulo Cortes and Pablo Altamirano. The name ALMA got ringing endorsements from Catherine Cesarsky (ESO Director General designate), who liked the fact that "alma" means "soul" in Spanish, and Ian Corbett, who liked acronyms he could pronounce. Later, after Japan joined ALMA, the acronym stood for Atacama Large Millimeter/submillimeter Array. Another important topic was the draft MOU for Phase 1 of ALMA, discussed earlier. It was initialed by Bob Eisenstein for NSF and H. Grage for the ENT to indicate approval by both parties

How ALMA Got Its Name

The origin of the ALMA name came from a random encounter and a decision with unexpected consequences. I was a newly minted engineer looking for what to do in life but without a clear idea of where to go. I also happened to have a thing for physics and astronomy but had no clear idea about how to become a professional astronomer. You see, in Chile undergraduate education is different. You do not get a major/minor; you have to go into a career path that will lead you to a professional degree, so you can get hired by someone. Because I didn't know what to do, I did physics and computer engineering so I could get a job. One day about March 1998, I was in the Information Technology Office at Universidad de Chile Astronomy Department, where I was doing some programming for a radio astronomer, when a scientist, Eduardo Hardy, entered with some questions. I didn't know who he was, but because my friend was busy, I engaged him in conversation. After thirty minutes he knew who I was and my field of expertise, and he asked me, *"What are your plans for the future?"* I said, *"I have no idea, I am applying to a job at Entel"* (a local communication company). He said, *"Why don't you work for us?"* Long story short, I made a fast decision and ended up at NRAO/Socorro. There I programmed data reduction software at the time when the MMA North American project joined with the European LSA project. A new name was needed (MMA and LSA are not that glamorous). An email had been sent asking for names. I started playing with words and two names came out: (1) ARTE for Atacama

Radio Telescope and (2) ALMA for Atacama Large Millimeter Array. Another Chilean engineer was there at the time, Pablo Altamirano. The two of us sent an email with the name suggestions. ALMA got chosen. Looking back, it was the right name, given the discoveries that ALMA has enabled. About a year later, I started graduate school to get a PhD in astronomy and ended up working as a scientist for the NRAO at the place that I named – ALMA – which is the unexpected consequence.

Paulo Cortes
Joint ALMA Observatory
Santiago, Chile

29 September 1999, Tokyo – The ACC first heard reports on progress in the United States, Europe, and Japan. A draft of an agreement making Japan a full and equal partner in ALMA was discussed. During a break in the agenda, a visit was paid to Monbusho, the Japanese science funding agency, by Catherine Cesarsky (by then ESO Director General), Ian Corbett, Bob Dickman, Riccardo Giacconi (by then AUI President), William Blanpied from the NSF office in Tokyo, Keiichi Kodaira, and Masato Ishiguro. Masayuki Inoue, the deputy director of Monbusho was not encouraging. Indeed, he complained about astronomers always asking for money. The visit was a classic example of differing perspectives and cultures: the ACC came with a viable two-party project to which they were willing and eager to add Japan. From their perspective, it would be a win-win for everyone. Inoue thought they simply wanted Japan's money. The ACC realized that funding constraints in Japan at the time would, in any case, preclude the signing of a tri-partite MOU. As an alternative, a resolution[14] was initialed that agreed to the establishment of a liaison between the ALMA and LMSA projects, with the goal of an eventual formal partnership in a millimeter/submillimeter array. To that end, the ACC formed the ALMA Liaison Group (ALG), chaired by Masato Ishiguro with Bob Brown as vice-chair. The ALG was charged with defining and evaluating the options for Japanese participation in ALMA that would be most scientifically and technically advantageous. Figure 6.5 shows a photograph taken of the ACC meeting participants.

12 November 1999, London – At this meeting, the resolution, initialed in Japan and signed there by Keiichi Kodaira, was also signed by Bob Eisenstein and Catherine Cesarsky. Reports were heard from the European and US project managers, including discussion of the purchase of prototype antennas. Bob Dickman reported the results of the Lehman Review. A report[15] on the first

Figure 6.5 Participants in the ACC meeting of 29 September 1999 in Tokyo. Standing, left to right: Makoto Inoue, Norbert König, Ryohei Kawabe, Ian Corbett, Bob Dickman, Paul Vanden Bout, Masato Ishiguro, Satoshi Yamamoto, Naomasa Nakai; seated, left to right: Catherine Cesarsky, Riccardo Giacconi, Keiichi Kodaira. Courtesy of Masato Ishiguro, reproduced by permission.

joint ALMA science symposium, held 7–8 October 1999 in Washington DC at the Carnegie Institution, was also presented. The ACC noted that the ALMA Science Advisory Committee (ASAC) had held an organizational meeting at the conference. The ASAC was charged[16] with refining the scientific goals of ALMA, and subsequently, with advising on the impact to those goals of proposed changes in technical specifications that might arise in the course of the project. A surprising development was the announcement from Masato Ishiguro that Monbusho, the Japanese science funding agency, had approved a draft tri-partite MOU.

7 April 2000, Washington DC – The most interesting action at this meeting, held at NSF, was in executive session and took up half of the time. Motions were quickly approved granting Chile an *ex officio* seat on the ACC and enlarging Japan's membership in the ASAC from three to five. This was followed by a contentious discussion of progress, or rather the lack thereof, in negotiations with Chile for a concession of the land for ALMA. Each side accused the other of failing to take the required steps. In fact, both sides were guilty. A proposal to let AUI take charge was rejected in favor of the appointment of a working group that would only be advisory to the ACC. The working group included Ian Corbett, Norbert König, Bob Brown, and Bob Dickman. Vanden Bout worried that involving a committee would only serve further to slow progress. The discussion then turned to the basis on which Japan would enter ALMA: providing additional capability, or merely saving the present partners costs. The former was preferred, but how? The AEC, defined earlier as a subset of the ACC, was charged with defining an array that optimized the science for $552 million (FY2000). In addition, the AEC was to work with the ASAC and the ALG to define a scientifically optimal enhanced project that included Japan, and to estimate its cost. The entrance of Japan into ALMA was becoming more serious, but was still a long way off.

13 October 2000, Paris – This meeting at the Observatoire de Paris had several routine reports. Dick Kurz and Marc Rafal presented a scheme for managing Phase 2, the construction of ALMA. They proposed organizing the project around IPTs, which had members from each side participating in the management of the execution of each ALMA delivery package. The scheme was well-known at NASA, ESA, and multi-national organizations. It did become the ALMA project management structure, with the incorporation of some quibbles over the organization chart. Then the meeting turned to Japan where national budget difficulties again made funding for Japan's participation in ALMA construction uncertain. Takayoshi Seiki, Director of the Research Institute Division of Monbusho made a strong statement[17] affirming the intent of Japan to join ALMA as a full and equal partner. He was followed by Norio Kaifu, the Director of NAOJ, who presented two options for Japan joining ALMA. The first involved adding another thirty-two 12 m antennas equipped with four receiver bands and an enhanced correlator. The second would add only fourteen 12 m antennas plus ten 8 m antennas with six receiver bands. Kaifu asked that an effort be made to conclude a tri-partite agreement by the following spring and that there be official Japanese members, not observers, in all the working groups. The latter request was granted, but, unfortunately, Riccardo Giacconi was somewhat abrupt in stating that any official participation of Japan in ALMA needed to wait until their effort was actually funded by the government. Understandably, Kaifu was offended. Vanden Bout recalls assuring him during a stroll in the Luxembourg Gardens that all would be well and Japan merely needed to be patient. Neither of them had even an inkling that it would take another four years of patience before funds were available and an MOU on the terms for Japan joining ALMA was concluded.

6 April 2001, Tokyo – This meeting[18] of the ACC marked a seemingly high point in the path of Japan's joining ALMA. The meeting began with a decision to add six Japanese members to the ACC, forming an Extended ALMA Coordinating Committee (EACC). Masayuki Shibata of MEXT[19] expressed strong support for Japan joining ALMA. He also noted the financial difficulties facing the Japanese government, but promised to work toward a start of the project. Yet another resolution was signed by Bob Eisenstein, Catherine Cesarsky, and Norio Kaifu committing to cooperative efforts to construct an agreement to bring Japan into ALMA, conclude negotiations with Chile for access to the Chajnantor site, and make best efforts to obtain approval and funding for the project. The resolution listed the members of the EACC. Following reports from Marc Rafal and Dick Kurz on progress in Phase 1, Catherine Cesarsky noted the common intent to buy antennas of one design. Masato Ishiguro pointed out that their prototype antenna was being developed by a partnership between the government and Japanese industry, and that purchasing a final single design could be problematic. This concern was echoed by Norio Kaifu. It was an early signal of the difficulties that would arise in contracting for identical 12 m antennas. Jack Welch presented a lengthy report on the recent meeting of the ASAC in Florence, Italy, in February 2001. He stated that the Japanese delegation was well integrated into the ASAC. In their second of 17 recommendations, the ASAC strongly supported the addition of the compact array that Japan proposed to contribute to ALMA in the event of a three-way partnership. Following the meeting, NRAO and ESO issued press releases announcing the imminent entrance of Japan into ALMA.

30 October 2001, Washington DC – It was at this meeting at NSF that the ACC learned from Ishiguro that the Japanese Finance Ministry would not be providing funds for ALMA in 2002, although funds for the purchase of a prototype antenna would be made available. It was expected that the Finance Ministry and MEXT would provide funds in 2004, but at a level only two-thirds that of the other two partners. The delay prompted a long discussion of how to proceed in the interim. It was concluded that it was important to continue the existing arrangements, if only to keep Kaifu in good standing with his ministries. Further, it was regarded as important to continue to define a tri-partite ALMA at the original level of financial participation by Japan. In other developments, Bob Eisenstein told the ACC that he was confident that there would be $12.5 million in the FY2002 NSF budget to start ALMA construction. His confidence was based on language that Senator Domenici of New Mexico had placed in the Senate appropriations bill. ESO had been slow in obtaining Council approval for funding, thinking that the chances for immediate funding in the United

States were still uncertain. Catherine Cesarsky asked Eisenstein to speak to the ESO Council at their December meeting.

19 April 2002, Venice – The Italian radio astronomers outdid themselves in the arrangements for the meeting of the ACC in Venice. The meeting took place in the ornate rooms of the Academy of the Lynx, founded in 1603 and the oldest science academy in the world. Gianni Tofani, a radio astronomer at the University of Bologna, was well connected in the art world and was able to schedule a private tour of San Marco Cathedral at night. But the attractions of Venice did not distract the participants from the serious issues of the meeting. Phase 1 was coming to a close and preparations needed to be made to manage Phase 2, and the actual construction of ALMA. Bob Brown presented an ALMA construction plan. Dick Kurz and Marc Rafal presented reports on the activities already begun on Phase 2 tasks. Eduardo Hardy reported on the negotiations with Chile regarding the site. At this point in the meeting, Bob Eisenstein spoke up. He was adamant that ALMA needed to be organized as a real project, with a director and project manager, reporting to a board. He was not prepared to support the funding of anything less. A project the size of ALMA, he argued, could not be run by a committee. No one disagreed but there was palpable tension in the room.

Brown and Kurz, after a quick whispered conversation, volunteered themselves as director (Brown) and project manager (Kurz). The ACC went into an executive session to discuss their offer. Both Riccardo Giacconi and Catherine Cesarsky were supportive of their offer, but the NSF representatives wanted others. The executive session ended with offers to Vanden Bout and Massimo Tarenghi to be (interim) ALMA director and project manager, respectively. The full meeting concluded with a declaration of intent to execute the Bilateral ALMA Agreement that was being negotiated by the end of the year.

17 September 2002, Garching – In the last three of four meetings held via telecon since their Venice meeting the ACC had: (1) made official the appointments of Vanden Bout (interim ALMA Director), Tarenghi (ALMA Project Manager), and Guilloteau (ALMA Project Scientist) effective 1 June 2002; (2) written job descriptions for these positions; and (3) appointed a search committee for Director and Project Engineer chaired by Anneila Sargent. She reported on the progress the committee had made. The meeting was memorable for those who were present in the executive session requested by Massimo Tarenghi. He presented a project organization chart that showed his impression of the project staff by highlighting their names in red, yellow, and green. Green managers were competent, yellow meant that with guidance they could become competent, and red was for those who should be fired immediately. He had tactfully omitted a color for Vanden Bout. Cesarsky mused in a stage whisper about the appropriate color for Tarenghi himself. No action was taken, but the ACC had been informed of his opinions as to where future trouble might lie.

29 October 2002, Santiago – Technically, this was not a meeting of the ACC, as the MOU establishing that body had expired. Nor was it the first meeting of the ALMA Board, as the Bilateral ALMA Agreement had yet to be approved and signed. But the group sensibly forged ahead with an "informal" meeting, which largely consisted of progress reports. Something significant to the future of the Joint ALMA Office (JAO) happened outside the meeting. After a few months on the job, Vanden Bout and Tarenghi decided they would only serve in their positions if they had full authority to exercise their responsibilities, with control of the budget, and without constantly checking with NRAO/AUI and ESO. This was conveyed to Cesarsky and Giacconi when the four met the evening before the meeting began, and there seemed to be agreement on the demand. But Giacconi and Cesarsky wanted to think on it, and the next morning they called Vanden Bout and Tarenghi's bluff. Institutional requirements trumped efficient management. The JAO would have responsibility, but little authority. Rather than quit, Vanden Bout and Tarenghi relented, the prospect of running the ALMA project trumped their management principles. From that point forward, JAO directors would not only need expertise in science and management, but would need whatever political and diplomatic skills it took to convince those holding the ALMA purse strings of the value of their plans. This arrangement may have been inevitable, however unfortunate. The JAO was not a legal entity. Only ESO and AUI could legally sign contracts and pay bills.

Following the execution of the Bilateral ALMA Agreement, the ACC simply morphed into the ALMA Board, changing from a committee into a board, and acquiring a new set of procedures that largely mimicked the old ones. The major change was that instead of a project run by a committee, ALMA had a well-defined management structure, exactly what Eisenstein had demanded. The ALMA organization chart for construction under the Bilateral ALMA Agreement is shown in Figure 6.6. The ALMA Parties, responsible for funding the project, were ESO and NSF/NRC for NA. The ALMA Executives, responsible for conducting the project, were ESO and AUI/NRAO. Project governance rested in the ALMA Board, advised by the ALMA Science (ASAC) and Management (AMAC) Advisory Committees. The JAO was responsible for execution of the ALMA Project Plan. Each area of the project was conducted by IPTs with memberships balanced between Europe and NA. In August/September 2004 ESO, NSF, and the National Institutes of Natural Sciences[20] (NINS) executed an agreement by which Japan officially joined ALMA, which added another arm to the central tree of the organization chart.

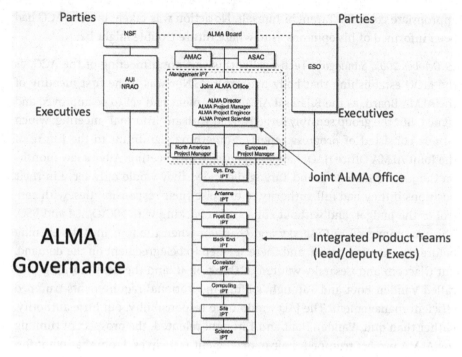

Figure 6.6 The organization chart for ALMA construction up to the point when Japan officially joined ALMA. Credit: NRAO/AUI/NSF, CC BY 3.0.

Musical Chairs

It might be expected that ALMA's long history would see changes in key personnel over the years. This was certainly the case. The most interesting change by far was the move Riccardo Giacconi made from being Director General of ESO to President of AUI. This happened in July 1999. He was succeeded at ESO by Catherine Cesarsky, a distinguished astronomer who had led the camera team for the European Space Agency's Infrared Space Observatory. This put the top position on each side in the hands of strong-willed management-savvy experienced hands. ALMA's success owes a great deal to their leadership. Giacconi's new associates at AUI and NRAO noticed his total, sudden switch in loyalty from Europe to the United States. It seemed as though it happened on his flight from Munich to Washington DC.

Up to the inauguration of ALMA in 2013, there were three ALMA directors – Paul Vanden Bout, followed by Massimo Tarenghi, and then Thijs de Graauw; three project managers – Massimo Tarenghi, followed by Tony Beasley, and then Dick Kurz; but only one project scientist – Richard Hills. John Credland served as the European project manager after Kurz. He was followed by Hans

Rykochevsky and then Wolfgang Wild. On the US side, the project managers were Marc Rafal followed by Adrian Russell. The European project scientists were Stepháne Guilloteau followed by Leonardo Testi. Al Wootten was the only US project scientist. For Japan, the project directors were Masato Ishiguro followed by Tetsuo Hasegawa, Ken'ichi Tatematsu, and, again, Tetsuo Hasegawa; project managers were Tetsuo Hasegawa, followed by Satoru Iguchi; project scientists included Ryohei Kawabe, followed by Koh-Ichiro Morita and Daisuki Iono and others; Satoru Iguchi and others served as project engineers. Although these changes could be distracting to upper management, at the working level life went on as usual. The project had a life of its own.

Notes

1 In 2006, T. Beasley negotiated a redesign of the points where the antennas crossed the pipeline.

2 A second pipeline from Argentina to Chile was constructed by NorAndino. It did not cross the Chajnantor site but took a nearby route. In 2016, it began transporting natural gas in the opposite direction – from a liquified natural gas port in Chile to Argentina.

3 As an example of the difficulties at the start of negotiations, Paul Vanden Bout and Jorge Molina Cárcamo, the governor of Region II where the ALMA site is located, got off to a bad start. The governor's monetary expectations from ALMA outraged Vanden Bout. Much later, at the ALMA inauguration, the two shook hands and celebrated the day.

4 A press release on the 10th anniversary of this agreement can be found at www.eso.org/public/austria/news/eso0621/.

5 *The MMA Program Plan – Design and Development Volume 2* can be found at: NAA-NRAO, MMA, MMA Planning, Box 7. https://science.nrao.edu/about/publications/alma.

6 The report of the Lehman Review can be found at: NAA-NRAO, MMA, MMA Planning, Box 9. https://science.nrao.edu/about/publications/alma.

7 The MOU defining the European Coordinating Committee (ECC) is Annex 1 of the document cited in note 8.

8 *ALMA Design and Development MOU Between NSF and Europe* (signed), December 1998. NAA-NRAO, ALMA, ALMA Multi-Institutional Agreements, Box 1. https://science.nrao.edu/about/publications/alma.

9 The initial ACC membership for the United States was: Bob Eisenstein, Hugh van Horn, Bob Dickman, Martha Haynes, Paul Vanden Bout, and a member of the US astronomical community to be named; for Europe: Riccardo Giacconi, Jan Bezemer, Ian Corbett, Arno Freytag, J-F. Minster, and Franco Pacini.

10 The LSA Board members were: Roy Booth (chair) (OSO), Harvey Butcher (NFRA), Marcello Felli (Obs. Arcetri), Michael Grewing (IRAM), Richard Hills (U. Cambridge), Jens Knude (Ast. Obs. Copenhagen), M. Mayer (Obs. Genève), Karl Menton (MPIfR), Peter Shaver (ESO), Jean Surdej (Inst. Astrophys. and Geophys., Liege), François Viallefond (Obs. Paris), and John Whiteoak (ATNF).

11 *The MMA/LSA Feasibility Study* can be found at NAA-NRAO, MMA, MMA Planning, Box 7. https://science.nrao.edu/about/publications/alma.

12 The ESO proposal can be found in NAA-NRAO, ALMA, ALMA Design and Construction, Box 2.

13 The JAO maintains a current technical description of ALMA for prospective users. For the version as of the writing of this book, see: https://almascience.nrao.edu/documents-and-tools/cycle10/alma-technical-handbook.

14 *Resolution between the ALMA* Coordination Committee and NAOJ Concerning Coordination between LMSA and ALMA (signed), 12 November 1999. NAA-NRAO, ALMA, ALMA Multi-Institutional Agreements, Box 1. https://science.nrao.edu/about/publications/alma.

15 The Washington DC conference in 1999 was the first of a series, to be followed by conferences held in Paris in 2004, Madrid in 2007, and Puerto Varas in 2012.

16 The fourteenth and final draft of the ASAC charter can be found at: www.cv.nrao .edu/~awootten/mmaimcal/asac/asac_charter.html.

17 The statement was included in the ALMA report of NRAO Newsletter #86, January 2001. https://library.nrao.edu/public/pubs/news/NRAO_NEWS_86.pdf.

18 The minutes of the ALMA Board meeting of 6 April 2001 in Tokyo can be found at: NAA-PVB, ALMA, ALMA: The Story of a Science Mega-Project. https://science.nrao.edu/about/publications/alma.

19 MEXT (the Ministry of Education, Culture, Sports, Science and Technology) was formed on 6 January 2001 from Monbusho (the Ministry of Education, Science, Sports and Culture) and the Science and Technology Agency.

20 The National Institutes of Natural Sciences (NINS) was formed in April 2004 as an inter-university research institute corporation consisting of five member institutes: the National Astronomical Observatory of Japan, the National Institute for Fusion Science, the National Institute for Basic Biology, the National Institute for Physiological Sciences, and the Institutes for Molecular Sciences.

7

Contentious Matters

"Necessity never made a good bargain."

Benjamin Franklin

The ALMA Board had to contend with a number of important and difficult issues which occupied much of the time in their meetings and the voluminous exchange of documents and emails that preceded the meetings. The partnership agreement between Europe and North America came up first. Next, the entrance of Japan into the ALMA partnership required long discussions over what Japan would bring to ALMA and the corresponding value of those contributions. Then there were the legal issues around the negotiations to secure the site. The procedure for the purchase of the 12 m antennas, the largest item in the budget, needed to be defined. Other sticky wickets to work through included the precise location of the ALMA headquarters in Chile and which of the partners would employ the local staff.

Bilateral ALMA Agreement

The MOU establishing the ACC was not a legally binding agreement. Even so, the NSF had agreed to let the ACC manage the ALMA activities in Phase 1 – design and development. But by the start of 2003, Phase 2 – construction – was about to begin. Adding to the urgency, as has been mentioned in the previous chapter, the MOU had expired at the end of 2002. Fortunately, the ACC had been busy for months refining draft agreements. If the final agreement was to be legally binding, it needed the blessing of the US State Department. Making the agreement an international treaty had been ruled out as that would have taken years to conclude. After review, the State Department pronounced the draft agreement as legally binding in an international court. The final draft was approved by the ALMA Board in its meeting on 24–25 February 2003 at NSF headquarters in Arlington, Virginia. On 25 February 2003, the Bilateral ALMA Agreement[1] was signed by representatives of the two partners: Rita Colwell, NSF Director, for NSF and the National Research Council (NRC) of Canada and Catherine Cesarsky, ESO Director General, for ESO, and Spain.[2]

The deliverables required of the parties to the Agreement were specified in the WBS that had been negotiated by Bob Brown, Masato Ishiguro, and Dick Kurz. The WBS was integral to the Agreement, as it committed the partners to delivering specific components. It included a schedule and total cost, as well as the management structure shown in the previous chapter as Figure 6.1. ALMA was to be completed by 2012 for a total cost of $552M (FY2000 US$) divided equally between Europe and North America. The distribution of the work packages went beyond establishing equal cost to the partners. The total risk was also balanced, giving each partner equal amounts of risk and contingency funds. It is ironic that the WBS, project schedule, and cost summary, all key elements of the Agreement, are in annexes. The body of the Agreement is largely concerned with rules of procedure, who owns ALMA, intellectual property rights, settlement of disputes, and the like.

Implementation of the WBS by the Executives would be a tremendously complex undertaking. As both mundane and challenging implementation matters arose, Pat Donahoe (AUI) and Ian Corbett (ESO) recognized that some sort of formal agreement between the Executives would be necessary to address a wide range of matters: protocols to be followed when interacting with Chilean authorities, issuance of contracts, relations with the Joint ALMA Office (JAO), the hiring of both local and international JAO staff, liability matters, common approaches to health, safety, security policies and procedures, and financial reporting. These and other matters were negotiated by Donahoe and Corbett and incorporated in a Bilateral ALMA Management Agreement,[3] signed on 31 July 2006 by Ethan Schreier, President of AUI, and Catherine Cesarsky, Director General of ESO. The management agreement proved essential to establishing commonly shared approaches to solving existing and future management challenges during ALMA construction and early operations. The Trilateral ALMA Agreement, discussed next, contains a requirement that all three Executives sign a similar management agreement covering operations.

Japan Joins ALMA

Following the 13 October 2000 ACC meeting in Paris, the Japanese government funded design and development work for the hardware contributions anticipated when Japan would eventually join ALMA. At first, the proposal was for Japan to become a full and equal partner. Accordingly, NAOJ contracted with the Mitsubishi Electric Co. (MELCO) for a 12 m prototype antenna. But it later developed that the funds Japan could provide would be approximately two-thirds of that required to match the US and ESO shares. This severely limited the number of 12 m antennas Japan could provide. However, the proposed compact array of twelve 7 m antennas, optimized for submillimeter operation, would provide a valuable and unique addition to ALMA. The offer to build additional receiver bands was also attractive, as ESO and US could only afford to build four bands out of the 10 that they had envisaged. The importance of having Japanese industry benefit from Japan joining ALMA played a role. For example, Japan proposed contributing a signal correlator for the compact array, presumably to be built by the Fujitsu Corporation, even though the correlator to be built for the large array had the capacity to handle all 66 antennas. The value of these options and the extent to which Japan would buy into the ALMA infrastructure were questions to be answered. As mentioned in the last chapter, the ALMA Board appointed a liaison group that held eight face-to-face meetings, with teleconferences in between, to arrive at a recommendation that satisfied the ASAC priorities and fit within the available funds. In August/September 2004, Arden Bement (NSF), Catherine Cesarsky (ESO), and Yoshiro Shimura (NINS) signed an agreement[4] formalizing Japan's entrance into ALMA, under Article 11 of the Bilateral ALMA Agreement, in exchange for what was called "Enhanced ALMA." Under the terms of the agreement, Japan would contribute $180 million (FY2000) to ALMA. The work packages included four 12 m antennas; a compact array of twelve 7 m antennas; equipping the entire array with Bands 4, 8, and 10 receivers; and a correlator for the compact array. This gave Japan a 25 percent share in ALMA. The agreement was amended twice, in June 2005 to extend deadlines,[5] and in June/July 2006 to recognize the collaboration between Japan and the ASIAA of Taiwan for the construction of Enhanced ALMA.[6] The ALMA Board also adopted an official statement describing ALMA. The most recent (2022) version of that statement is illustrative of the complexity of ALMA:

> The Atacama Large Millimeter/submillimeter Array (ALMA), an international astronomy facility, is a partnership of the European Organisation for Astronomical Research in the Southern Hemisphere (ESO), the U.S. National Science Foundation (NSF) and the National Institutes of Natural Sciences (NINS) of Japan in cooperation with the Republic of Chile. ALMA is funded by ESO on behalf of its Member States, by NSF in cooperation with the National Research Council of Canada (NRC) and the Ministry of Science and Technology (MOST) in Taiwan and by NINS in cooperation with the Academia Sinica (AS)

in Taiwan and the Korea Astronomy and Space Science Institute (KASI). ALMA construction and operations are led by ESO on behalf of its Member States; by the National Radio Astronomy Observatory (NRAO), managed by Associated Universities, Inc. (AUI), on behalf of North America; and by the National Astronomical Observatory of Japan (NAOJ) on behalf of East Asia. The Joint ALMA Observatory (JAO) provides the unified leadership and management of the construction, commissioning, and operation of ALMA.

Figure 7.1 In the front (from left to right): Tim de Zeeuw (ESO Director General), Katsuhiko Sato (NINS President), and F. Fleming Crim (NSF Assistant Director) on behalf of France Córdova (NSF Director) representing the three partners at the ceremony and signing the Trilateral ALMA Agreement for Operations. In the middle row (from left to right): Phil Puxley (Program Director, NSF), Masahiko Hayashi (Director General of NAOJ), Hideyuki Kobayashi (Deputy Director of NAOJ), Satoru Iguchi (East Asia ALMA Project Manager, NAOJ), Yuko Nagasaka (Translator, NAOJ). In the back row: Jim Ulvestad (Division Director, NSF), Rob Ivison (ESO Director for Science), Junichi Watanabe (Deputy Director General of NAOJ), Tsuyoshi Sasaki (Head of Administration, NAOJ), Shin-Ichiro Asayama (NAOJ Chile), Nikolaj Gube (Deputy Head Legal and International Affairs, ESO), and Takao IIzawa (Head of Administration, NINS). Credit: ALMA/ESO/AUI/NINS, CC BY 4.0.

The Trilateral ALMA Agreement[7] between ESO, NSF, and the NINS was signed in Tokyo, Japan, on 15 December 2015. The new agreement provided the framework for the long-term *operation* of ALMA over the next 20 years. The meeting attendees posed for a photograph, shown in Figure 7.1, to record the event. It had taken more than 12 years since the signing of the Bilateral ALMA Agreement to reach this point.[8] The Management Agreement previously concluded between ESO and AUI was updated in the Trilateral ALMA Management Agreement[9] signed on 17 November 2016 by Ethan Schreier, AUI President; Tim de Zeeuw, ESO Director General; and Masahiko Hayashi, NAOJ Director General.

Wandering the Road from Nobeyama to ALMA

The road from Nobeyama to ALMA was not straight. After working 10 years with solar radio interferometers at Toyokawa Observatory, I moved to the Tokyo Astronomical Observatory of the University of Tokyo in 1980 and started working as a leader of the construction of the Nobeyama Millimeter Array (NMA). I started the design work at Toyokawa before moving to the Nobeyama Radio Observatory (NRO). Since I had no experience in non-solar radio observing and aperture synthesis, the design work was completely new to me.

In November 1972, I had attended the Symposium on the *Collection and Analysis of Astrophysical Data* held at NRAO Charlottesville. After the meeting, I visited various observatories in the world to see the radio interferometer systems at Stanford University, Caltech Owens Valley, NRAO Green Bank, Mullard Observatory (UK), and Westerbork Observatory (The Netherlands). Travelling alone around the world helped me gain an international perspective in my life. The insights I gained from this visit were applied to the design of the NMA. The first aperture synthesis observations with NMA were made at 22 and 115 GHz in 1984 and 1988, respectively. Around 1983, I realized that a larger millimeter wave interferometer would be needed in the future, and began to study the basic concept of the Large Millimeter Array (LMA), which later became the Large Millimeter Submillimeter Array (LMSA).

While I was the NRO Director, between 1990 and 1996, I worked very hard to make the LMSA plan a reality. As a first step, in 1992, I visited Chilean sites for the first time with Naomasa Nakai and Ryohei Kawabe in collaboration with SEST/ESO. The same year I organized IAU Colloquium 140 at Hakone, Japan, with Jack Welch as co-chair, to discuss the status of the operating millimeter and submillimeter arrays and the future plans for larger arrays. At this conference, I sensed a great deal of interest in future international cooperation in larger arrays. With a tailwind for international cooperation, Roy Booth and I proposed the establishment of the Millimeter-Submillimeter Array Working Group at the Kyoto URSI General Assembly in 1993. These activities were followed by lively discussions of cooperation between MMA and LMSA starting in 1994.

The most shocking event in my life with ALMA was the signing of the MOU toward a joint project between NRAO and ESO. It happened in June 1997 when I visited NRAO Charlottesville to discuss the antenna design with Peter Napier and Jeff Kingsley. Paul Vanden Bout, then NRAO Director, invited me to his office and told me that the MOU was signed on that day by him and Riccardo Giacconi. It was very surprising message for me, because I was told beforehand by Bob Brown that an official signing would not happen in the meeting. For me, it was like having my bride stolen before the wedding. Anyway, after various twists and turns, the projects of the three parties, Japan, North America, and Europe, were finally unified to ALMA.

In 2007, I decided to step down from the ALMA-J Director and joined the JAO as a lead test scientist to work for the Assembly-Integration-Verification (AIV) group. At that time, I was the first and only Japanese staff member. I could work with many Chilean engineers under the leadership of Joe McMullin. Working with the Chilean staff as well as the US and European staff was a great pleasure for me and made my 16-month stay in Chile very enriching. In February 2009, I returned to Tokyo for my retirement from NAOJ. As I was very eager to see the first fringes from ALMA, I decided to visit Chile again in November 2009. During my visit, I was attacked by a high fever and cough and was admitted to a hospital in Calama. After staying at the hospital three days, my medical condition miraculously recovered and I was able to return to work at ALMA. I saw the first fringes at Band 9 just one day before I left Chile.

I feel very happy that I was involved in various aspects of the ALMA project, such as initial planning of LMSA, site surveying, budget request preparation, international negotiations, and the AIV work in Chile (Figure 7.2).

Masato Ishiguro
Professor Emeritus, National Astronomical Observatory of Japan
Tokyo, Japan

Figure 7.2 Masato Ishiguro with some members of the AIV team at his going-away party. Left to right: Joe McMullin, Dick Sramek, Fernando Gallardo, Masato Ishiguro, Johnny Reveco, Cristián Maureira, Camilo Menay. Credit: M. Ishiguro, used by permission.

Site Negotiations

The process of obtaining a concession to the Chajnantor site was discussed in the previous chapter. Here we address some legal issues that occupied the ALMA Executives, still only two at the time when the site was negotiated. Both were eager to obtain a concession from the Chilean government that allowed the construction and operation of ALMA on the Chajnantor site. At the same time, both were concerned with the legal terms and conditions that a concession agreement might entail. The legal standings of ESO and AUI in Chile were and still are very different, and the concession would need to be consistent with both without compromising either. ESO is an international treaty organization, similar to CERN or the United Nations. Its Convenio with Chile is based on that of the United Nations Economic Commission for Latin America and the Caribbean. ESO interacts with Chile at the level of the Ministry of Foreign Affairs. ESO has an ambassador called the ESO Representative to Chile. Not compromising its existing legal standing was critical to ESO, with its substantial investments in the La Silla and Paranal Observatories. AURA was also watching from the sidelines, concerned that its agreement with Chile for the Cerro Tololo Inter-American Observatory (CTIO) not be adversely impacted by the ALMA negotiations. The same was true for the Carnegie Institution of Washington and its Las Campanas Observatory (LCO). At that stage, AUI was established in Chile on the same basis as AURA and Carnegie, through an agreement with U. Chile under Law 15172.

As already noted, the slow pace of the negotiations for the site was of sufficient concern to the ALMA Board that a working group was appointed to expedite matters. It was intended to provide better communication with the board and allow individual group members to undertake coordinated assignments. For example, arrangements were made to have the German ambassador to Chile urge the Ministry of Foreign Affairs to act, and for the American ambassador to Chile to speak to the governor of Chile's Region II, where the ALMA site is located. There was plenty of time to review the proposed terms of the agreements with the agencies of the Chilean government that were involved, and the board provided substantial feedback to the negotiators. Daniel Hofstadt and Eduardo Hardy, ESO and AUI Representatives in Chile, respectively, were critical in pushing the process along. In the end, the Ministry of Foreign Affairs decided that the ESO Convenio did not need to be replaced. It could cover ALMA in addition to the other ESO observatories, and an agreement to that effect was signed. AUI was granted the same rights and privileges as ESO under Law 15172. The entire process was finally concluded in late 2003 in a contract with BN giving the ALMA Executives the use of the ALMA site for 50 years.

Contracting for the Antennas

The 12 m antennas that made up most of the array constituted the largest item in the ALMA budget, estimated to be $228 million or 41 percent of the original budget cited in the Bilateral ALMA Agreement. Accordingly, the process of selecting a contractor to provide these antennas was very carefully constructed. The intent of the North American and European partners was to select a single design from a single contractor for the production of the entire array of 64 antennas, choosing the design that met the specifications at the best price. A motivation for a single design was that an array with multiple antenna designs incurs greater maintenance costs: more spare parts, increased training for antenna mechanics, multiple antenna control systems, and the like. In addition, different antenna designs could introduce different systematic errors into specific observations, such as polarization. Having a single design was a laudable goal, but as it turned out, one doomed to failure.

Specifications – As a first step in antenna procurement, bids were solicited for prototype antennas which would be tested to verify their compliance with the specifications. To meet ALMA's science goals, the 12 m antennas needed to go beyond the state-of-the-art. No antennas of this size and precision had ever been built. The reflecting surface needed not only to be of exquisite accuracy, but also to maintain that accuracy as the antenna changed direction on the sky. The ability to point in a given direction with very high accuracy, and to change pointing direction quickly, was needed as well. The resulting specifications addressing these points and many more were developed over many years, beginning with those for the MMA, LSA, and LMSA.

The key specifications were:

- Reflector Surface Accuracy: 25 µm with 20 µm goal
- Absolute Pointing Accuracy: 2 arcsec over the entire sky
- Offset Pointing Accuracy: 0.6 arcseconds within a 2 degree radius
- Fast Switching: 1.5 degree move in 1.5 s, with a 3 arcseconds peak pointing error
- Path Length Accuracy (Non-Repeatable): 15 µm
- Path Length Accuracy (Repeatable): 20 µm

   These specifications are to be satisfied under all azimuth and elevation angles of the antenna and all environmental situations, in particular, wind of 6 m/s by day and 9 m/s at night, as well as full solar illumination from changing directions.

To illustrate two of the above key specifications, consider that an antenna was required with a 12 m diameter reflecting surface that deviates from a perfect parabola by no more than 20 micrometers (µm), which is about one-third the diameter of an average human hair or about the size of a human white blood cell, averaged over the surface. This specification allows observing at frequencies up to 950 GHz (0.32 mm in wavelength). The antennas must also be able to point by dead reckoning to any direction on the sky with an accuracy of 2 arcseconds, approximately 1/14th of the smallest angle that can be perceived by the human eye.

NRAO/AUI received several proposals on 30 June 1999 to design and build a prototype. The proposal evaluation was done using standard procedures. A Contract Selection Committee (CSC) was appointed, chaired by Bob Brown. Business and Technical Evaluation Committees ranked the proposals on their respective merits, without any knowledge of likely cost, and reported their findings to the CSC. The CSC then made a final ranking and opened the price.

The proposal from Vertex[10] was judged by the CSC to be superior to the others. ESO and NAOJ followed their own procurement procedures, in the end selecting different contractors. North America purchased a prototype antenna from Vertex, Europe from the European Industrial Engineering Consortium[11] (EIE), and Japan from the MELCO. The contract with Vertex with a price of $6.1 million was signed on 21/22 February 2000. The fact that the Vertex contract[12] is over 200 pages long indicates the detail and specificity needed to state the very large number of requirements. The purchase orders for all three prototypes contained identical technical specifications.

Prototype Antenna Testing – The three contractors delivered their prototypes, shown in Figure 7.3, to NRAO's VLA site in New Mexico for testing. There the testing team could use the extensive VLA infrastructure to advantage. The VLA site is only 2,120 m above sea level, compared to 5,000 m for the ALMA site, but it is sufficiently high and dry much of the year for testing of an antenna at modest (3 and 1.2 mm) wavelengths. On the other hand, wind at the VLA site and its poor optical "seeing" (the atmospheric effect that makes stars twinkle) would make the tests of antenna pointing accuracy difficult. The Vertex prototype was first to be available for testing in March 2003. Testing of the prototypes from Vertex and EIE was done by the Antenna Evaluation Group (AEG), a joint working group of engineers and antenna experts from Europe and North America who were appointed to test both

Figure 7.3 The ALMA prototype antennas (left to right): Japan's, North America's, and Europe's. Credit: NRAO/AUI/NSF, CC BY 3.0.

the Vertex and EIE antennas. The AEG began with the Vertex antenna immediately after its assembly. The EIE prototype, carried to the Albuquerque airport in an Airbus Beluga, could not be tested by the AEG until January 2004. The MELCO prototype arrived in May 2003 and was assembled by a small army of engineers and technicians by September 2003. Japan conducted tests of their MELCO prototype separately, as Japan was not yet an official partner in ALMA. The testing program was to have been completed by 1 June 2004, but the late delivery of the antennas, especially the EIE antenna, forced testing of the antennas to run later.

The AEG had the opportunity to test and compare two prototypes with very different designs. Vertex had designed an antenna that pushed conventional techniques to the limit by incorporating carbon-fiber-reinforced plastic (CFRP) in a few key elements. EIE used far more CFRP. Both designs had insulated steel bases and yokes (the "forks" that support the elevation "tipping" structure). But EIE built the receiver cabin at the antenna's secondary focus entirely from CFRP, whereas Vertex made theirs from insulated steel. The reflecting panels for both antennas are supported by a Backup Structure (BUS) made from CFRP. Vertex connected the cabin to the BUS with an Invar[13] ring to limit thermal expansion, whereas EIE used CFRP throughout. The reflecting panels were also very different in the two designs. Vertex made machined aluminum panels, etched to produce a surface that would scatter solar radiation and prevent overheating at the secondary focus. EIE electroformed nickel skins on accurate molds, glued them to thick aluminum honeycomb cores, and coated them with a thin layer of rhodium to scatter solar radiation. The prototypes had radically different drive systems: Vertex used a conventional gear drive and EIE a direct drive system with permanent magnet linear motors. The resulting EIE design was lighter at 80 metric tons than the Vertex design at 108 metric tons.[14] The tests pitted an upgraded conventional design from Vertex against an *avant garde* design from EIE. Which would be better?

A major AEG task was adjusting the reflecting surface panels to form a parabola with the specified accuracy of 20 μm. As delivered, the surfaces were smooth to 80 μm, four times worse than the desired accuracy. Topographic images of the surfaces, obtained by a holographic technique, showed where adjustments in the surface panels needed to be made. By repetitively iterating between making slight adjustments and follow-up holographic images, the surfaces could be set to the final accuracy. Both prototypes satisfied the specification. The same was true for the pointing accuracy, both prototypes meeting this requirement, as well as for many other tests. One design was occasionally slightly better in some specifications and the other in other specifications, but both designs appeared to satisfy the ALMA requirements. The AEG issued a report[15] on 28 May 2004. A complete description[16] of the testing program and results was published later by the AEG leadership. A report of the results[17] of the tests by the Japanese team showed that the MELCO design met the key ALMA specifications, although the testing program given in the report appears to be somewhat less extensive than that of the other two prototypes. All three of the prototypes seemed to be qualified for the ALMA large array. One nagging issue remained that would lead to more testing. The structure of the Vertex prototype did not deform under changing gravitational loads as it should have according to computer models.[18]

Prototype Destinies – After the Japanese astronomers learned that their funding to join ALMA would be less than what was hoped for, it was clear that only four 12 m antennas could be afforded in addition to the compact array of 7 m antennas, new receiver bands, and correlator. The prototype would be one of these four and MELCO would manufacture three more. The North American partners judged the cost of moving the Vertex prototype to Chile to be more than that of a new antenna. They had no other viable use for it and NSF declared it to be surplus property. Proposals were solicited by NSF from would-be owners. Two proposals were received. One was from a team at the University of Arizona, led by Lucy Ziurys, a leading astrochemist. Had they won, the Vertex prototype would have replaced the old NRAO 12 Meter Telescope, which Arizona had acquired as surplus property in 2000. But the alternate proposal was selected. It came from a partnership between the ASIAA of Taiwan and the Smithsonian Astrophysical Observatory, with principal investigators Paul Ho and Roger Brissenden, respectively. They planned to place the antenna first at Thule Air Force Base and later at the NSF's Summit Station on the high ice sheet (3,200 m elevation) of Greenland. That would make it the northernmost millimeter wavelength telescope on the planet, at an ideal location for a future element of the Event Horizon Telescope (EHT), which would later image the super massive black holes at the center of the galaxy M87 and our own Milky Way. Initially disappointed, the Arizona team went on to acquire the EIE prototype, which was moved to Kitt Peak from the VLA site in 2013, where it continues to operate today.

Production Antennas – Even though the testing of the prototype antennas was not complete, on 17 December 2003, NRAO/AUI and ESO issued a call for proposals to build up to 32 antennas each. Each Executive's call contained identical statements of work which were jointly developed in conjunction with the JAO. Each Executive had its own business terms and conditions which were required by their respective laws, policies, and procedures, and which would also give rise to pricing differences. The proposals were due on 30 April 2004 with pricing valid until 31 October 2004. The goal was to procure antennas of a single design that satisfied the performance specifications at the lowest possible price. The joint process established to review proposals and select a contractor is shown in Figure 7.4. The process is similar to that followed by NRAO for selecting a prototype antenna contractor, but it takes into account the different contract terms and conditions between AUI/NRAO and ESO. The Proposal Receipt Teams (PRT) open the proposals, secure and seal the prices, and confirm that the proposals are generally responsive. Then the proposals, without prices, are sent to the Joint Technical Evaluation Team (JTET) and in the case of NRAO/AUI, the Business Evaluation Team. The results of JTET evaluation were sent to the ESO Contract Award Committee (CAC) and to the AUI/NRAO CSC which also received a report from the Business Evaluation Committee (BEC). After that step, the standard procedures of each side are followed.

The overall goal of both contracting processes, as confirmed by the ALMA Board on several occasions, was for each side to procure the same antenna design. However, after extensive, and sometimes intense, discussion between the Executives and by the ALMA Board over the preceding years, the process also allowed for the possibility of acquiring different designs.

ESO limited its call for tender (CfT) to *qualified European* companies, where *qualified* meant the company had successfully passed ESO's Preliminary Enquiry Process. NRAO/AUI did not require that proposals came from US companies. However, NRAO/AUI did limit participation by the following language in the RFP:

> This solicitation specifically reserves the right to award the production antennas to the successful prototype Contractor. Participation in the AUI antenna procurement is limited to those Proposers/entities that have made a substantial contribution to either the AUI or the ESO prototype antenna.

Proposals were received from Vertex, and a consortium[19] consisting of Alcatel Space France, European Industrial Engineering S.r.L., MAN (AEM) on the US side, and from the European subsidiary of Vertex – Vertex Antennentechnik (VA), Alenia Space Italy, and AEM on the European side, consistent with the above requirements. The proposals were opened on 3 May 2004 in a teleconference between ESO, NRAO/AUI, and the ALMA Director. At this point, an enormous effort[20] began that involved nearly 30 meetings and negotiations on the US side alone. The proposal receipt committees checked that the proposals were responsive, sealed the prices, and passed the proposals to the JTET, and on the US side to the BEC, for review. The JTET and Antenna Technical Working Group (ATWG) reports[21] went to the Contract Selection/Assessment Committees. After extensive study, the NRAO/AUI CSC gave both proposals under their review comparable scores with respect to specifications and general business

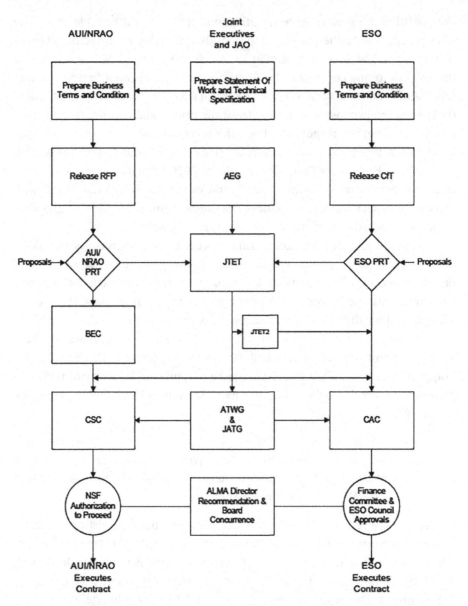

Figure 7.4 The flow chart for joint evaluation of antenna production proposals and the selection of a contractor. The acronyms are: Request for Proposal (RFP), Call for Tender (CfT), Antenna Evaluation Group (AEG), Proposal Receipt Team (PRT), Business Evaluation Committee (BEC), Joint Technical Evaluation Team (JTET), Contract Selection Committee (CSC), Contract Award Committee (CAC) Antenna Technical Working Group (ATWG), and Joint Antenna Technical Group (JATG). Credit: NRAO/AUI/NSF, CC BY 3.0.

qualifications. However, in addition to a significantly higher price, the CSC noted that AEM proposal had several shortcomings: instead of a fixed price, the price was indexed for inflation without a cap; the proposal did not contain a design for the specified common transporter interface; and AEM made numerous exceptions to NRAO's contract terms and conditions. AEM was informed on 29 August 2004 in a letter[22] from Bill Porter, NRAO Business Manager, that they would no longer be considered by the US side unless negotiations with another vendor failed. In a further exchange of letters, AEM confirmed the continued validity of its price through 31 October 2004.

Although the CSC had judged the Vertex proposal to be superior to AEM's, the review had raised a number of questions to be asked of Vertex. These were addressed in a series of three meetings between Vertex and NRAO: on 27–28 July 2004, to discuss a myriad of proposal details; on 18–19 August 2004, to discuss more proposal details and pricing; and on 6 October 2004 to review and understand the revised offer pricing submitted by Vertex on 8 September 2004. Even though the price was the lowest received in a competitive bidding process, it was more than double the amount budgeted for antennas in the Bilateral ALMA Agreement. The main reason for the difference was the dramatic increases in the prices of steel and oil, the latter being the raw material for CFRP. Changes in the design of some elements and the cost of providing an assembly building, lodging, meals, and utilities at the OSF added to the increase as well.

To ensure careful coordination of the antenna procurements on both sides of the Atlantic, a series of four meetings were held over the summer and autumn of 2004 at the Heathrow (7 July and 24 August 2004) and Dulles (2–3 August and 29 November 2004) airports. These were high-level meetings with the Director General of ESO, the President of AUI, and the NRAO Director in attendance. The meetings also included the ALMA Director and ALMA Project Manager. The first three meetings were routine, with each side reporting to the other on its plans and activities. But the fourth meeting was intensely confrontational. The ATWG could not certify from their examination of the AEG prototype test results that the Vertex antenna would meet specifications.[23] As has been mentioned, the concern was the failure of the structural computer model provided by Vertex to accurately predict the measured deformation of the antenna under varying gravitational loads. Bob Dickman pressed the group to contract with Vertex for the production antennas anyway, motivated by the rapidly rising cost of commodities and confident that the technical issues would be solved. In fact, NRAO/AUI had *already* requested NSF permission to do this on 14 September 2004. The ESO delegation was strongly opposed to Dickman's proposal. They considered ignoring the lack of conclusive test results to be wildly imprudent. The meeting nearly fell apart. In the end, Dickman relented and it was agreed that a new testing campaign would be conducted. A new team, the JATG, would review the previous testing and make new tests as required to resolve the technical issues. The JATG team comprised participants in the earlier testing plus additional experts.

The main problem was eventually traced to an error in the holography software. Although smooth to the required accuracy, the surface was not set to a perfect parabola but one that had a superimposed "donut" shape. The error was discovered by Robert Lucas[24] from IRAM in examining the results of a simple test. The distortion was just barely detectable in the photogrammetry measurements that had been made previously. Once holographic measurements and surface setting had been accomplished with the software error corrected, a consistent set of all the test results was obtained. This testing took time and the report was not given to the ALMA Board until April 2005. The final report[25] of the JATG certified both prototypes as meeting specifications:

> Based on all available data and the ATF testing done by the AEG and the JATG, it is the consensus view of the JATG that both prototype antennas meet the ALMA antenna specifications under direct consideration (surface accuracy at all elevations, all-sky absolute pointing performance) under the environmental conditions encountered during the testing, and that the production antennas based on either design can also be expected to meet these specifications.

Based on this information, the NRAO/AUI CSC recommended that authorization again be sought from NSF to execute a contract with Vertex. Rapidly rising commodity costs were driving the price of the antennas up and there was heavy pressure to buy the antennas as quickly as possible. The authorization[26] to purchase was sought from NSF on 15 April 2005. The request provided pricing covering the common antenna design as well as the scenario under which ESO would procure its antennas of a different design. The request, as modified on 5 May 2005 in response to some NSF questions, put the NSF into an extremely difficult position. The results of the re-baselining effort to assess the cost to complete ALMA, discussed in the next chapter, were not yet in, but it was already clear that a substantial increase in the ALMA budget would be required if the project was to continue. NSF was far from identifying the additional funding required to support the escalating costs of the project.

At ESO the process had apparently come to the same conclusion. As a result of contract selection coordination activities between AUI/NRAO and ESO prior to, and in the margins of, the ALMA Board meeting of 5–6 April 2005 in Pasadena, California, it was becoming clear that ESO was moving toward choosing VA, the German subsidiary of Vertex. At the meeting, Cesarsky told the Board that she planned[27] to submit the Contract Award Committee's recommendation to the Finance Committee (FC) by 15 April and that if her proposal was approved at the 10–11 May meeting of the FC, it could be ratified by the ESO Council at its 6–7 June meeting. At its May meeting, the FC approved the selection of Vertex Antennentechnik, and authorized ESO to begin negotiations. However, as it turned out, the FC also decided[28] it needed assurance from Council that ESO could afford to pay for its entire share of the ALMA project.

The ESO Council was scheduled to affirm the affordability of ALMA in its 6–7 June meeting in Helsinki, but the German delegation would not support the resolution without evidence that the costs would not escalate even further. The German position was supported by a majority of the Council members.[29] The recommendation to the FC to approve the purchase of antennas from VA had been greeted with consternation by some of the Council members. Most importantly, national interests were expressed in the meeting by the Italian delegation. They complained about the way the procurement was being conducted. At one point in the meeting, there was a threat to take Italy out of ESO if the contract did not go to AEM.[30] Indeed, Piet van der Kruit, Council President, recalled "*I was in a very sad mood and was very concerned that the whole ALMA project would come to an end.*" The meeting ended by passing a resolution confirming ESO's commitment to ALMA and that it would purchase its antennas on a sound financial basis.

The 21 June 2005 ALMA Board meeting at Den Haag, Netherlands was a crucial point for the ALMA project. In addition to Dickman who was the NSF representative, NSF also sent to the meeting Wayne Van Citters, Astronomy Division Director, Judy Sunley, MPS Deputy Assistant Director, and Mike Turner, the NSF Assistant Director for MPS, upon whose desk AUI/NRAO's antenna approval request was sitting.

Turner believed in what he called "transformational science." By that he meant science that transformed a discipline, opening up new avenues and directions of research. Turner was convinced that ALMA would produce transformational science. Waiting until the anticipated cost reviews were complete would result in even higher antenna costs. A pre-publication copy of a report[31] from the National Academies of Science, Committee on Astronomy and Astrophysics had been transmitted to the NSF in June 2005. It concluded that although none of the three Level-1 Science Goals could be reached with a 50-antenna array, transformational science was still possible and the project deserved continued NSF support. If an increased project budget were not approved, the project would die and the contract would be canceled. But if increased funding were realized, all would be well. He was persuasive in arguing the case with NSF upper management.

At the meeting, the North American representatives presented two resolutions from the AUI Board of Trustees: one which urged NSF to approve the AUI contract for the North American antennas and another in which they welcomed *"ESO Council's reaffirmation of its strategic commitment to ALMA and encouraged ESO along with the ALMA Director and the ALMA Board to take all necessary steps to ensure the success of the project."* Massimo Tarenghi, the ALMA Director, urged the Board to endorse the US purchase of antennas to avoid further delay and cost increase. During a break in the ALMA Board proceedings, Turner and Sunley met with Ethan Schreier (AUI President), Fred Lo (NRAO Director), Anneila Sargent (AUI and ALMA Board member), and Pat Donahoe (AUI Secretary) and after quickly assessing the risks and rewards of authorizing the procurement Turner made what he termed *"a bold move"* to authorize the NSF ALMA Board representative to vote in favor of the motion to proceed with the award of the AUI/NRAO contract with Vertex. The ALMA Board then reconvened and gave its formal approval.[32] On 11 July, AUI/NRAO signed a contract with Vertex. The results of the JAO's revised project budget would be available in September and it was hoped that ESO could proceed to a purchase of VA antennas soon after that. It looked like ALMA would have a homogeneous array of antennas. But this was not to be.

The FC and ESO Council decisions to effectively delay the award of the VA contract meant that ESO had several choices[33] with respect to their CfT, the principal ones being: terminate the CfT and do nothing, issue a totally new CfT, or simply extend the existing CfT for as long as the end of the calendar year. The first choice would kill the ALMA partnership and was unthinkable. ESO chose to ask their vendors to extend the pricing validity of their proposals until 30 September 2005, but did not specify a date by which the vendors could submit any further revisions to their proposals, a fact that would come into play later.

From the beginning, there had been considerable pressure on ESO to select the AEM proposal. As early as late September of 2004, the previous year, Cesarsky had been invited to a meeting[34] with the Italian Minister of Science in which he urged her to sign a contract with the French-Italian consortium. On 28 October 2004, letters[35] had been received from the Italian Minster of Education, Universities, and Research, and from Piero Benvenuti, the head of the Istituto Nazionale di Astrofisica and senior spokesman of the Italian astronomical community. Both letters expressed concern on the process of the procurement. As noted above, in the meantime, the Italian members of Council had continued to question the validity of the selection process, and AEM had begun to send fax messages to Council members and to attempt to reach them by phone.[36] In the face of all this and to her great credit, Cesarsky was resolute in her commitment to buy a single design.

Regrouping, van der Kruit, Cesarsky, and Corbett envisioned a plan[37] which consisted of holding a meeting of the ESO Committee of Council (Council's "executive committee" – CoC) on 16 September 2005 at which the CoC would discuss the status of the antenna procurement, and then a 29 September 2005 combination CoC/Council meeting to address the affordability determination that had been requested by the FC in its May meeting. If approved by the Council, then Cesarsky would send her recommendation to the Finance Committee for approval at its October meeting.

In preparation for this sequence of events, ESO sought proposal clarifications from VA and the AEM consortium. On Tuesday, 13 September 2005, VA submitted bid clarifications to ESO of its firm fixed price and business terms and conditions, and pointed out that although their current proposal was already compliant with ESO's requirements regarding the amount of work that was to be done in Europe, it would be working later in the month to determine whether this allocation could be further improved. VA also extended its price validity to 31 October 2005. On Thursday, September 15, the AEM Consortium submitted to ESO clarifications of its firm fixed price bid and extended the validity of its price to 31 October 2005. On Friday, 16 September, the CoC met and the contract situation was discussed at length. Then came a surprise – on Tuesday, 20 September, AEM submitted another revised firm fixed price proposal which significantly reduced their previous pricing. The timing of the revised bid was striking. As the President of the ESO Council at the time put it,[38] "*It is not clear why this occurred at this particular time, but it did happen after the Committee of Council meeting where it was stated that if things would remain as they stood, Vertex would win.*" The next day ESO declared[39] that the bidding process had ended and Cesarsky sent her AEM award recommendation to the Finance Committee for consideration at its upcoming October meeting. The AEM bid was now lower than the bid from VA and Cesarsky had no choice but to recommend that AEM receive the contract.

In the meantime, on 8 September 2005 the JAO had released its estimate of the cost to complete the project, upon which Cesarsky had prepared a paper that demonstrated the affordability of ALMA, consistent with ESO's long-range plans, assuming Spain would join ESO.[40] At the 29 September 2005 combined CoC/ESO Council meeting the paper was discussed and thus the affordability question that had been initially raised by the FC in May. The Council then affirmed the affordability of ALMA,[41] and also directed the FC to continue with its proceedings with respect to awarding the antenna contract. On 5 October 2005, the FC approved the negotiation and award of a contract to the AEM Consortium.[42]

This decision to award the contract to AEM instead of VA was a totally unexpected result, at least to most outside observers. Ever since, there has been speculation on both sides of the Atlantic that confidential information was leaked to AEM. Although it certainly fits the circumstances, we can cite no evidence that it happened. The Vertex price for the US antenna purchase was public. AEM could have modified that price for European contract requirements and other calculable differences and then submitted a bid that was lower. Perhaps "Euro-fever" ran high at AEM and they submitted a very low bid on which they were willing to take a loss. The mystery remains. ESO signed a contract with AEM on 7 December 2005 for the purchase of 25 antennas.[43]

Following the FC's antenna decision, on two occasions, VA wrote to ESO[44] raising what VA viewed as violations of fundamental formalities in the procurement process including ESO's lack of establishing a common deadline for final price submissions. In late October 2005 Ethan Schreier, AUI President, received a letter[45] from VA which cited the procedures that ESO had followed and proposed that ESO should hold a public bid opening to once and for all eliminate any perception of possible misconduct. Similar letters were sent to NSF. VA also proposed that AUI vote against the award of the contract at the upcoming ALMA Board meeting. Schreier discussed the situation with Turner, who, while sharing his disappointment in the ESO decision, was also extremely concerned about the upcoming external cost review of the JAO's estimate to complete the project, the result of which might pose an existential threat to ALMA. In the end, NSF and AUI concluded that voting against the award of AEM contract had the potential of further destabilizing the project. At the November 2005 meeting in Santiago, Ian Corbett took pains to review[46] the actions at ESO in great detail, the implication being that, following a completely above-board process, ESO was compelled to take the lower priced bid. It was in all probability true, but some in the United States delegation were left with an uneasy feeling that in the future they should be wary in their dealings with their counterparts at ESO. The ALMA Board also extensively discussed the ESO decision's implications of proceeding with two antenna designs. In the end, the Board concurred with the ALMA Director's recommendation that ESO proceed with the issuance of the antenna contract.

NSF commissioned an ALMA Delta Cost Review (ADCR) to assess the additional costs resulting from the selection of two different antenna designs. It met on 26 January 2006 at NSF and concluded[47] that the cost to the project of a mixed array of 50 antennas was $8 million more than the original budget for a homogeneous array. Relative to the total budget for antennas, this was a small amount, but it did represent the price of about one antenna. However, this was exclusive of the increased antenna costs to the United States that came from Vertex spreading its non-recurring costs over 25 antennas rather than 50. The hit on operations was estimated to be a 10 percent increase. These extra costs, seemingly significant in themselves, were not that large compared to the over-all budget and the huge scientific payoff that was to come.

The original Vertex contract price was indexed for inflation and capped at $183,000,000 prior to any modifications. The cost of the Vertex antennas after all modifications was $183,113,889, an increase of less than one-tenth of one percent.[48] The final cost of the AEM contract was never released. Contrary to the widespread fear that an array of two different antenna designs would be costly to ALMA operations, in ten years of experience the expense has proved to be much less than the ten percent estimated by the ADCR.[49] The impact on sci-entific results, specifically in the area of polarization observations, has turned out to be well managed by the data analysis software. However, at the time, the decision to buy a second design for the European antennas took a toll on trust between certain individuals at ESO and NRAO/AUI. There were two contentious issues yet to be resolved: the location of the JAO and the employer of the JAO's local staff. Fortunately, the shared goal of realizing ALMA overcame personal differences.

Local Staff Employment

Every JAO employee is legally employed by one of the ALMA Executives, either ESO, NAOJ, or AUI. The JAO staff take programmatic direction from the ALMA Director while matters of compensation and other terms of employment remain with the legal employer. The JAO staff is divided into two categories: Local Staff (LS) and International Staff (IS). These categories are not original to ALMA; they have always been used by ESO, AURA, and Carnegie. As origi-nally implemented, an IS position was one requiring job skills that were not available in the Chilean labor market. By agreement among the Executives, IS positions were allocated to ESO, NAOJ, or AUI depending on the nature of the position, budget availability, and the desire to achieve an equitable balance of such positions among the Executives. The resolution of who would be the legal employer of the LS, constituting the overwhelming majority of JAO staff, would take years to resolve. A wide range of options was considered, including having each Executive hire its *pro rata* share. This was judged to be too complicated and it was decided that there should be a single employer. Would that be ESO or AUI? (This issue arose before Japan joined ALMA; Japan showed no interest in employing the LS after it joined.)

From the outset, ESO had made it clear that it did not want to hire the LS, but rather proposed a third alternative – outsourcing the employment of the LS. Under this arrangement ESO and AUI would form a jointly owned Chilean corporation to hire the employees and handle labor relations. It would *per force* be subject to Chilean labor law and the jurisdiction of Chilean courts. From AUI's perspective, this hiring arrangement posed several problems, chief among them was that such an arrangement could serve to undermine the status of the other US observatories in Chile, namely CTIO and LCO, specifically, that it might lead to curtailing their immunities. ESO also proposed that the LS be outsourced to an existing Chilean company, one that would not be owned by AUI and ESO. However, AUI had substantial reservations about whether outsourcing the LS employment to a totally independent company might be viewed by Chilean authorities as an illegal arrangement. AUI's position was that the only satisfactory hiring arrangement would be that AUI hire the LS. Following the practice at AURA and Carnegie, AUI would follow Chilean labor law even though it was not obligated to do so.

After years of haggling over details, consulting lawyers, and exchanging letters and emails, ESO concluded that their proposal was going nowhere. Immediately after ESO announced its decision on the antenna contract, Ian Corbett and Pat Donahoe met[50] privately in Chicago in early October 2005 to develop a process that would, once and for all, decide how the LS would be hired. They agreed to utilize the Disputes Provision of the ALMA Agreement to elevate the matter to the NSF Director and the ESO Council President. On 10 January 2006, ESO Council President, Richard Wade, accompanied by Monnik Desmeth and Laurent Vigroux, met at NSF in Arlington, Virginia with NSF's Mike Turner, representing the NSF Director, Wayne Van Citters, and Judy Sunley to settle the matter. NSF and ESO acknowledged the complexity of issues surrounding the matter, and that each side had genuine concerns about each of the Executives' preferred options. At the conclusion of the meeting, a recommendation was made to have AUI hire all LS for the JAO. The result was not a surprise to either AUI or ESO because Corbett and Donahoe had done extensive backchanneling over previous months to achieve this outcome. AUI was directed to prepare an ALMA Local Staff Implementation Plan. The plan[51] was approved by the ALMA Board at their June 2006 meeting in Santiago.

Unions are strong in Chile, particularly so in the north where there is a long history of union activism in the mining industry. The vast majority of ALMA workers are unionized. Initial relations between the ALMA union and AUI were good and a contract was successfully negotiated in 2010. In 2013, the union did call a strike and ALMA ceased operations for a brief period until an agreement on a new contract was reached. Since then, labor relations have been less confrontational, with contract renewals in 2015, 2018, and 2020 reached without strikes. Worker compensation in Chile is periodically indexed for inflation. As a result, union demands are more focused on working conditions and benefits. In 2019 a second union was organized, mostly comprised of employees in Santiago and managers and supervisors at the ALMA site. AUI has never invoked its immunities, instead choosing to follow Chilean labor law and appearing in labor court when sued. As a result, ALMA enjoys overall good labor relations that prevail at CTIO and LCO, as NSF hoped would happen. An outstanding example of ALMA's commitment to its employees occurred during the height of the COVID-19 pandemic when everyone was paid even when no one was allowed to come to work and astronomical observing was shut down.

At the time of this writing in 2022, there were 214 LS employees. The IS employees were distributed as follows: 11 working for AUI, 16 for ESO, and eight for Japan. As new skills became available in the Chilean labor market, Chilean applicants began to be awarded IS positions, and six had been successful by 2022. Also working on ALMA are the staffs of the Executives in Santiago: 20 at AUI, 10 in the Japanese office, and a similar number at ESO.

ALMA Headquarters Location

The location of the JAO, also known as ALMA's Santiago Central Office (SCO), was an equally thorny issue to resolve. It made economic sense to build it on land available in ESO's Santiago facility,[52] not to mention its location in an attractive, upscale area of Santiago called Vitacura. However, AUI was eager to display itself in Chile as an equal ALMA partner with ESO and feared that locating the JAO at ESO would make ALMA look like just another ESO observatory. Riccardo Giacconi, AUI's president, was particularly opposed to a location at ESO's headquarters in Chile. AUI explored several alternatives to the ESO location, including contacting U. Chile regarding locating the building on the Cerro Calan Observatory site on the east side of Santiago.

In an 11 May 2004 letter[53] from the ESO Director General, Catherine Cesarsky to Bob Dickman at NSF, ESO formally offered to construct the SCO building on ESO's Vitacura site. Conscious of NSF's and AUI's views about ALMA being viewed as an ESO initiative, rather than an international collaboration, the letter specifically stated that, while a specific site within the ESO property could not be identified in the absence of a definitive design, the building would have "... *an address and entrance separate to those of ESO buildings.*" Little progress was made toward a definitive decision about the building's location until the 6–8 April 2005 meeting of the ALMA Board when it endorsed[54] a paper jointly prepared by AUI, ESO, and the JAO which outlined a series of actions to be undertaken based on the premise that the ESO proposal of the Vitacura location was the preferred alternative. The ALMA Director was tasked to undertake actions which would result in a final decision which would be subject to separate ALMA Board approval. However, no more progress occurred during the next six months largely due to the activity surrounding the antenna procurement decisions and the hiring arrangement for JAO local staff.

Events which would result in a decision picked up momentum as a result of the 10 January 2006 meeting between NSF and ESO Council representatives who not only recommended that AUI should hire the JAO LS but also recognized the benefits of the offer of land at the Vitacura site for the location of the Santiago building. At its 22–24 March 2006 meeting, in Kyoto the ALMA Board officially recognized the tradeoff made by the NSF/ESO Council delegation between the ALMA building site and LS employment, and approved the selection of ESO's Vitacura site as the location for the ALMA Headquarters building, noting that implementation should be consistent with ESO's 11 May 2004 letter to the NSF.[55] As a result of joint efforts to identify specific locations within the ESO site, three potential locations were identified. Two locations were on the north side of the property away from ESO's main entry gate, with the third location immediately adjacent to the main entry gate. From AUI's and NSF's perspectives, the least preferred location was by the ESO main gate because it blurred the important point that ALMA not be viewed as an ESO enterprise, but rather as an international collaboration. Nevertheless, in November 2006, ESO Council approved the site by the main entrance as the "*best location*" for the ALMA building. This decision was met with considerable dismay by NSF and AUI because they thought that ESO Council failed to appreciate the depth of their concern. Over the next year as the detailed building design process evolved, ESO stated that it was impractical to establish a separate entrance. On 8 March 2007, NSF Director Arden Bement formally accepted the offer of the Vitacura site in a letter[56] noting that, "*ESO, the JAO, and NSF are working together to secure architectural designs consistent with the need for ALMA to have an image in Chile as a separate international collaboration with its own distinct identity.*" The conditions in ESO's 11 May 2004 offer letter to NSF citing that the facility would have a separate entrance and address were never fulfilled. But on the bigger issues of economics, convenience, and surroundings, the location of the building is a success. The SCO is shown in Figure 7.5.

Figure 7.5 A view from the north of the SCO (ALMA Headquarters) on the ESO compound in Vitacura, Santiago. The lawn separates SCO from the ESO building, out of the image to the right. The small building to the right leads to an underground parking garage between the two buildings. Credit: ESO, CC BY 4.0.

Notes

1 A draft of the Bilateral ALMA Agreement can be found at NAA-NRAO, ALMA, ALMA Multi-Institutional Agreements, Box 3. https://science.nrao.edu/about/publications/alma.

2 Spain would join ESO on 16 February 2006.

3 The Bilateral ALMA Management Agreement can be found at NAA-NRAO, ALMA, ALMA Multi-Institutional Agreements, Box 4. https://science.nrao.edu/about/publications/alma.

4 The agreement bringing Japan into the ALMA partnership can be found at: NAA-NRAO, ALMA, ALMA Multi-Institutional Agreements. https://science.nrao.edu/about/publications/alma.

5 The first amendment can be found at NAA-NRAO, ALMA, ALMA Multi-Institutional Agreements. https://science.nrao.edu/about/publications/alma.

6 The second amendment can be found at NAA-NRAO, ALMA, ALMA Multi-Institutional Agreements. https://science.nrao.edu/about/publications/alma.

7 The Trilateral ALMA Agreement can be found at NAA-NRAO, ALMA, ALMA Multi-Institutional Agreements. https://science.nrao.edu/about/publications/alma.

8 For a history of Japanese millimeter interferometry from NRO to ALMA, see Ishiguro, Chiba, and Sakamoto (2022).

9 The Trilateral ALMA Management Agreement can be found at NAA-NRAO, ALMA, ALMA Multi-Institutional Agreements. https://science.nrao.edu/about/publications/alma.

10 The ownership history of Vertex is complex. CPI Vertex Antennentechnik GmbH originated as the Antenna Division of Friedrich Krupp AG in 1968. Krupp AG, together with MAN, built the Effelsburg 100 Meter Telescope. Its counterpart in the United States, the 100 m Robert C. Byrd Green Bank Telescope, was built by Radiation Systems Inc. (RSI). In 1994, RSI was acquired by COMSAT, Inc., which was in turn acquired by Lockheed Martin in 1999. In 2000, RSI was sold by Lockheed Martin to Tripoint Global, a company held by the Thyssen-Bornemisza Group, which had acquired Vertex Antennentechnik (VA) in 1999. In 2004, General Dynamics bought both companies from Tripoint Global and merged RSI and VA into a division called Vertex Satcom Technologies. In 2019, long after ALMA construction was complete, Vertex Satcom was acquired by CPI, Inc. We ignore the series of buyouts and takeovers in this account, referring to the US company as Vertex and using VA for its European subsidiary.

11 ESO's contract for the purchase of a prototype antenna was originally placed with a consortium consisting of European Industrial Engineering, S.r.l. and Costamasnaga. Following the bankruptcy of Costamasnaga, the contract was amended in December 2003 by ESO, recognizing the reorganization of the consortium with Alcatel Space as the leader. We refer to the prototype contractor as EIE.

12 The purchase order for the Vertex prototype antenna can be found at NAA-NRAO, ALMA, ALMA Design and Construction, Box 1. https://science.nrao.edu/about/publications/alma.

13 Invar is a steel alloy with a high content of nickel and a very low coefficient of thermal expansion.

14 The Vertex and AEM (EIE) production antennas weighed 97.7 and 89.5 metric tons, respectively, both meeting the limit set in the antenna specifications of 100 metric tons.

15 *Evaluation of the ALMA Prototype Antennas Report (Executive Summary)*, dated 28 May 2004, can be found at NAA-AUI, Projects, ALMA. https://science.nrao.edu/about/publications/alma.

16 Mangum, J.G. et al. (2006) give a full report on the North American/European prototype testing program and its results.

17 Ukita, N. et al. (2004) present the results of the tests of the Japanese prototype.

18 Mangum, J.G. et al. (2006). See the Executive Summary, Figure 6.13.

19 The Alcatel Space France, European Industrial Engineering S.r.L., MAN (AEM) consortium changed in July 2005 to become Alcatel Alenia Space France, European Industrial Engineering S.r.L., MT Aerospace. This occurred when the MAN Technologie subsidiary of MAN was sold to form MT Aerospace. This happened after the JTET report of 15 June 2004; the acronym remained the same.

20 The US process of proposal review leading to the selection of a production antenna contractor is described in the ALMA Antenna Procurement Summary of 15 April 2005. It can be found at NAA-AUI, Projects, ALMA, Box 5. https://science.nrao.edu/about/publications/alma.

21 The JTET report can be found at NAA-AUI, Projects, ALMA. https://science.nrao.edu/about/publications/alma.

22 Porter's letter to AEM can be found at NAA-PVB, ALMA, ALMA: The Story of a Science Mega-Project. https://science.nrao.edu/about/publications/alma.

23 The ATWG reports of 29 September can be found at NAA-NRAO, ALMA, ALMA Advisory
    and Coordinating Committees, Box 4. https://science.nrao.edu/about/publications/alma.
    The addendum of 17 November 2004 can be found at NAA-PVB, ALMA, ALMA: The Story
    of a Science Mega-Project. https://science.nrao.edu/about/publications/alma.

24 A memorandum by Robert Lucas reporting his discovery of the error in the holography
    software can be found at NAA-PVB, ALMA, ALMA: The Story of a Science Mega-Project.
    https://science.nrao.edu/about/publications/alma. A thorough description of the
    holography technique, including the proper near-field correction, and its application to
    the ALMA prototype antennas was published by Baars et al. (2007).

25 *ALMA Joint Antenna Technical Group (JATG) Test Results, 2005* can be found at NAA-AUI,
    Projects, ALMA, Box 5. https://science.nrao.edu/about/publications/alma.

26 The letter to NSF requesting approval of the purchase of production antennas from
    Vertex can be found at NAA-AUI, Projects, ALMA, Box 5. https://science.nrao.edu/about/
    publications/alma.

27 Our account of the production antenna procurement events at ESO draws heavily from
    the memoir of Piet van der Kruit, *Five and a Half Years in ESO Council*, pp. 77–95. It can be
    found at NAA-PVB, ALMA, ALMA: The Story of a Science Mega-Project. https://science
    .nrao.edu/about/publications/alma.

28 Ibid., p. 87.

29 Ibid., p. 88.

30 Ibid., p. 88.

31 The report is available at http://nap.naptionalacademies.org/11326.

32 P. van der Kruit, *Five and a Half Years in ESO Council*, p. 93.

33 Corbett's memorandum, *Antenna Procurement – Options, 7 June 2005*, can be found at NAA-
    PVB, ALMA, ALMA: The Story of a Science Mega-Project. https://science.nrao.edu/about/
    publications/alma.

34 P. van der Kruit, *Five and a Half Years in ESO Council*, p. 78.

35 Ibid., p. 86.

36 Ibid., p. 91.

37 Ibid., p. 94.

38 Ibid., p. 95.

39 NAA-NRAO, ALMA, ALMA Board Meetings.

40 P. van der Kruit, *Five and a Half Years in ESO Council*, p. 94.

41 A copy of the Council resolution can be found at NAA-PVB, ALMA Series, ALMA: The Story
    of a Science Mega-Project. www.nrao.edu/archives/items/show/38359.

42 P. van der Kruit, *Five and a Half Years in ESO Council*, p. 94.

43 Ibid., p. 94

44 The letter VA sent to ESO can be found at NAA-PVB, ALMA, ALMA: The Story of a Science
    Mega-Project. https://science.nrao.edu/about/publications/alma.

45 The letter to E. Schreier from R. Vardeman can be found at NAA-PVB, ALMA, ALMA: The
    Story of a Science Mega-Project. https://science.nrao.edu/about/publications/alma.

46 The minutes of the ALMA Board meeting of 5 November 2005 can be found at NAA-
    NRAO, ALMA, ALMA Board Meetings.

47 The ALMA Delta Cost Review Report can be found at NAA-PVB, ALMA, ALMA: The Story of
    a Science Mega-Project. https://science.nrao.edu/about/publications/alma.

48 The total cost of the North American 12 m antennas can be found in the release of claims located at NAA-PVB, ALMA, ALMA: The Story of a Science Mega-Project. https://science.nrao.edu/about/publications/alma.

49 Beasley to Vanden Bout, private communication.

50 Donahoe to Vanden Bout, private communication.

51 The ALMA Local Staff Implementation Plan can be found at NAA-NRAO, ALMA, ALMA Board Meetings.

52 A report on the options and costs of various locations in Santiago for the JAO offices was prepared by the JAO. It can be found at https://library.nrao.edu/public/memos/alma/misc/ALMAU_2.pdf.

53 Cesarsky's letter to Dickman can be found at: NAA-PVB, ALMA, ALMA: The Story of a Science Mega-Project. https://science.nrao.edu/about/publications/alma.

54 The minutes of the ALMA Board meeting 6-8 April 2005 can be found at: NAA-NRAO, ALMA, ALMA Board Meetings.

55 The minutes of the ALMA Board meeting 22-24 March 2006 can be found at NAA-NRAO, ALMA, ALMA Board Meetings.

56 Bement's letter can be found at: NAA-PVB, ALMA, ALMA: The Story of a Science Mega-Project. https://science.nrao.edu/about/publications/alma.

8

Funding ALMA

"Success in research needs ... luck, patience, skill, and money."

Attributed to Paul Ehrlich

What was true for the great German pharmacologist Paul Ehrlich in the late nineteenth century remains true today, not only for research itself, but for research facilities as well. The funds that made ALMA possible were the result of very different processes in Europe, East Asia, and North America. Each process required good fortune, patience, and skill in varying measures depending on the circumstances. Fortunately for astronomy, governments around the world were ultimately generous in their support, sharing a vision for transformational science.

Europe

That the European participation in ALMA was united under ESO made for a huge simplification of the funding and approval process. The alternative would have required negotiations with at least a half dozen science funding agencies who, in turn, would need to persuade their governments to supply the money. It was the approach first taken by NRAO when trying to find international partners – specifically in Europe – for the MMA. Luckily, that effort was not successful, doomed by the feeling in Europe that a large millimeter wavelength array should only be built as a pan-European project. ESO exists to do precisely that.

The procedure at ESO for approving and thereby funding a major project is superficially simple. It only requires the consent of the ESO Council. The Council is made up of two representatives from each member state, one of whom is from the member state's government, so that their vote of approval reflects their government's commitment to provide funds. ESO's budget is guaranteed, made up of contributions from the member states that are a percentage of each state's gross domestic product. ESO's budget can vary with economic conditions in Europe, but continuity from year to year is assured. The largest projects, which cannot be accommodated in that basic budget, require special contributions over and above the annual payments. If sufficient commitments can be obtained, and if the project has been shown to be scientifically meritorious and technically feasible by the Scientific Technical Committee (STC), the project proposal goes to the ESO Council for approval. It should also be mentioned that ESO can borrow money if necessary. Its assured annual funding makes it a very low risk borrower.

Even though ESO Council has the ultimate say, a process of study and review every bit as elaborate for ALMA as those in the United States, Canada, and Japan was undertaken and lasted for several years before the final consent was given, following the important endorsement of the STC. In some sense, the timing of ALMA as a potential ESO project was ideal. This did not escape the notice of Riccardo Giacconi, ESO Director General at the time, who was a savvy strategic thinker. The Very Large Telescope (VLT) on Paranal was nearing completion. ESO's budget to build the VLT was large, and if continued at that level could fund the next big project. Plans for an Overwhelmingly Large Telescope (OWL) were underway, but that project was not shovel-ready. ALMA was a scientifically exciting project that could fill the gap. The timing was also ideal in that the United Kingdom wished to join ESO. When a new state becomes a member of ESO it must pay an amount equal to its share, based on gross domestic product, of the existing ESO infrastructure. In the case of the UK, this initiation fee prevented the need for any increases in the special contributions from other member states as ESO pivoted from the VLT to ALMA. The proposal to build ALMA was submitted to the ESO Council in December 2000. An operations plan followed in March 2001. The ESO Council approved the proposal in July 2002.

Japan

Progress to funding Japan's participation in ALMA might have been hampered by the changes in bureaucratic contacts that accompanied the absorption of Monbusho into MEXT as the ministry responsible for science and technology and the later creation of NINS. However, Norio Kaifu, the NAOJ Director, had the political skills to manage these shifting sands. Progress was definitely set back by the economic crisis of 2001, following a tripling of prices for stocks and real estate in Japan in the years prior to 2001.

When the Bank of Japan increased interest rates, the bubble in asset prices burst, and along with it the dream of the Japanese radio astronomers to be full and equal partners in ALMA. The decade 2001–2011 has been called the "lost decade," an era in which the Japanese economy stopped growing and the government found itself with tight budgets. Gradually, the mood changed and by 2004 MEXT was willing to commit to $180 million (FY2000) to be spent over eight years, completing Japanese deliverables on the same schedule as ESO and NRAO.

The proposal for Japan to join ALMA had to undergo rigorous review. The Council for Science Technology of MEXT was one such major hurdle. Final review was by the Council for Science, Technology, and Innovation, a cabinet-level body that included among its members no less than the prime minister, Jun-Ichirō Koizumi. In August–September 2004 an agreement was signed by Arden Bement (NSF), Catherine Cesarsky (ESO), and Yoshiro Shimura (NINS) that made Japan a partner in ALMA. Shimura was a constant supporter of Japanese radio astronomers through all the reorganizations of funding agencies. The agreement was amended in 2006 to recognize an agreement between Taiwan's ASIAA and NINS. Taiwan joined forces with Japan just as Canada had with the United States, making the ALMA parties Europe, North America, and the East Asia. The patience of the Japanese radio astronomers had paid off. The agreement was announced in the NRAO Newsletter[1] by Al Wootten.

> Japan entered the ALMA Project with the signing of an Agreement Concerning the Construction of the Enhanced Atacama Large Millimeter/Submillimeter Array. The Agreement was signed by Arden L. Bement, Acting Director of NSF, Catherine Cesarsky, Director General of ESO, and Yoshiro Shimura, President of the National Institutes of Natural Science (NINS) of Japan. All parties had signed the document by September 14, 2004, making Japan an official partner in an Enhanced ALMA, to be known as the Atacama Large Millimeter/Submillimeter Array (same ALMA acronym). Final negotiations on an operations plan for Enhanced ALMA are expected to be concluded by the end of 2005. Japan will provide the Atacama Compact Array with its correlator, three receiver bands, and other components. The value assigned to the Japanese contribution to Enhanced ALMA is $180M (FY2000 US$). Assuming all three partners are able to meet their commitments, the final project will be cost-shared 37.5 percent, 37.5 percent, 25 percent, between North America, Europe and Japan, respectively. The observing time, after a ten percent share for Chile, will be distributed accordingly.

North America

A new Congress of the United States is convened every two years follow-ing elections, and passes legislation that, among many other things, authorizes the expenditure of government funds. What a given Congress cannot do is bind a future Congress to actually appropriate the necessary funds for projects that have been authorized. This stricture means the annual budgets of a multiyear project like ALMA are subject to review and appropriation every year. Although appropriations committees often follow the guidance in an agency's budget request, they are not obligated to do so and continuity is not guaranteed. This fact of the US science funding process makes it far more political than those of ESO or Japan. Consequently, good fortune, patience, and skill played large roles in achieving funding for the MMA and the US participation in ALMA.

The MMA had received the first funds of a three-year program of design and development in January 1998, while Neal Lane was the NSF director. It was expected that funds for construction of the MMA would follow in NSF's Major Research Equipment (MRE) line. In August 1998 Lane was succeeded as director by Rita Colwell, a prominent biologist. In the NSF budget request for FY2001, National Ecological Observatory Network (NEON) appeared in the MRE line and the MMA was notably absent. As FY2000 was to be the third and final year of MMA design and development, it was not clear what would happen to the project if it failed to gain funds for construction. Putting the construction of the MMA behind that of NEON posed an existential threat to the MMA and US participation in ALMA.

NSF, as part of the executive branch of the US government, is forbidden by law from lobbying Congress. The legislative side of its Office of Legislative and Public Affairs (OLPA) only works, in principle, to maintain good relations with the relevant congressional committees, answer their questions, and keep them informed. Other visitors to the "Hill" (the cluster of congressional offices surrounding and including the Capital), it was hoped by NSF, would only argue for support for science in general, as did the American Astronomical Society and other professional science organizations on a regular basis. Advocating for particular NSF programs was discouraged, with an important exception. If a project was approved at NSF and included in its budget request, then out-side appeals to Congress for the project were not only sanctioned but quietly encouraged. Because the NSF had funded design and development, the engage-ment of professional help to secure Congressional support for ALMA construc-tion was appropriate.

From its beginning, AUI had considered hiring a legislative liaison firm as being unnecessary. But when Riccardo Giacconi became AUI President on 1 July 1999 that changed. Perhaps from his experience as Director of the

Space Telescope Science Institute, he felt that advancing ALMA in the United States Congress required professional expertise and experience. At the urging of Anneila Sargent, he decided to engage Lewis-Burke Associates, a firm specializing in science advocacy. Sargent knew the head of the firm, April Burke, who handled legislative affairs for Caltech. The AUI Board of Trustees was of mixed opinion. Those who had worked in university administrations tended to be supportive while the more idealistic faculty members were opposed. But Giacconi, fresh on the job, was given approval to go ahead. He accompanied Burke on only one visit to a congressman's office, that of Representative James Sensenbrenner of Wisconsin, the chairman of the House Science Committee. The visit was a bad experience for Giacconi, indeed, it was so uncomfortable for him that he said he would never do it again. In future, he said, Vanden Bout would make the visits. Lewis-Burke would turn out to be an invaluable aid to funding ALMA. Hiring them was one of Giacconi's best decisions at AUI.

NSF budget requests are sent to the Office of Management and Budget early in the government's fiscal year, which starts on 1 October, for inclusion in the President's budget request for the following fiscal year. The request is embargoed until the President's budget request is sent to Congress. The situation for the MMA – that there was no request for MMA construction, or even for more design and development, for FY2001 – was known to all interested parties by the time of the VLA 20th anniversary celebration on 23 August 2000. The event was significant in that both Senator Domenici of New Mexico, ranking minority leader on the Senate Budget Committee, and NSF Director Colwell had accepted invitations to attend and make remarks. The acceptance was with some trepidation on Colwell's part. NRAO had submitted a proposal to upgrade the VLA on 1 July 2000. She had been warned by OLPA[2] that the senator would be sure to urge her to support the project. She was willing to do this, but only if the funding was additional to NSF's budget request. While NRAO would have welcomed Domenici's support of the Expanded Very Large Array (EVLA) project, staving off the impending death of the MMA had higher priority.

Ultimately, that rescue had to wait a year. Domenici was proud of the presence of the VLA in his state and it came ahead of the MMA in his priorities. In earlier visits to his office,[3] the conversation would always turn from the MMA to the VLA. While no one but the two participants know what was said between Domenici and Colwell in the private meeting they had during the anniversary celebration, the Senate report language[4] accompanying the NSF appropriation for FY2001 made the sense of their meeting clear:

> In last year's Senate report, the Committee expressed its support for enhanced operations and maintenance and development of new instrumentation at the Very Large Array and the Very Long Baseline Array in New Mexico ...

> *these astronomy facilities need to be supported in their operations, and new instrumentation and upgrades must be provided toe keep them as world class facilities. Accordingly, the Committee provides an additional $13,000,000 above the fiscal year 2001 request levels for the astronomical sciences subactivity for these facilities. ... The Committee recommends an appropriation of $109,100,000 for major research equipment. ... The Committee has provided ... $6,000,000 for the Millimeter Array.*

The EVLA Project would be started with a portion of the extra funds added to AST budget and the MMA would receive the funds needed to continue for another year of design and development. Dickman recalls being notified of the design and development funding as a *"correction to a misunderstanding in the Director's office that had zeroed out additional funding."* NRAO made an estimate of what a one-year delay in the start of construction would cost. But what would happen in FY2002? Would ALMA construction begin then?

NSF's intention became known in the spring of 2001 when yet again the budget request for the MRE line did not include the MMA. Once again, the project faced a crisis. There was no request for construction funds or even a continuation of design and development. Would Senator Domenici help this time? Now that the EVLA project was underway, he turned out to be more amenable to discussing ALMA.[5] The EVLA and ALMA were both large radio interferometers, making it possible, in principle, to construct electronic components that would serve both projects. It was stated in visits to Domenici's office that although ALMA was to be located in Chile, much of the electronics would be made by NRAO in Socorro.[6] At the same time, at the urging of Lewis-Burke Associates, a letter campaign was organized. A list of 86 radio astronomers was asked to write letters[7] to their Congress members urging support for ALMA. While there is no direct evidence of the campaign's impact, Domenici did save the day. The Senate language for the FY2002 NSF budget[8] provided the first construction funds for ALMA in the amount of $12.5 million.[9]

> *The Committee recommends an appropriation of $108,832,000 for major research equipment. ... The Committee has provided ... $12,500,000 for initial construction of the Atacama Large Millimeter Array (ALMA) radio telescope. ... The Committee supports initiation of construction of the ALMA radio telescope and has provided the necessary resources to start construction.*

In November 2001, the National Science Board approved the start of ALMA construction at the request of NSF Director Colwell. This was a huge step, essentially guaranteeing US partnership in ALMA. The following year went according to plan: the NSF requested $30 million for ALMA construction, a request that was funded by Congress. The US participation in ALMA would not have happened were it not for the fact that a powerful Senator chose to give it his support simply because his state was home to an NRAO facility in which he took pride. In two years, ALMA would face another near-death crisis, this time one of global dimensions involving all the partners.

Project Reset

In 2004, it became clear that the cost estimate for the project was significantly less than what was needed. During the two years since construction began, funds budgeted for contingency were used to make up for increased costs. After Massimo Tarenghi became Project Director, the search for his replacement as Project Manager led to hiring Tony Beasley, who had just completed the construction of CARMA. Beasley accepted the position, moved to Chile, and discovered that he had signed on to a project in trouble. Unless drastic action was taken, funds available from contingency would soon be exhausted and ALMA would run out of money well short of project completion. The project needed to be "rebaselined," that is, it needed a new credible cost to complete, hopefully on the same schedule.

The process for establishing a new budget was similar to that of the first cost estimate, but this time with the benefit of experience. Every element of the WBS was reexamined and an estimate of the cost of the labor and materials to complete it was made, down to the individual part. A risk (the odds of failure) associated with completing the task was assigned and used to calculate the contingency that should be added to the estimate. For example, a task with only a 10 percent chance of failure would have a contingency of 10 percent of the estimated cost. Then the estimates and contingency for all the elements were added together. A cost estimate for the entire project was completed by 8 September 2005. Table 8.1 shows a comparison between that estimate and the original budget. The new project estimate, with contingency, implied that $242 million (FY2000) in additional funding was needed to complete ALMA. This was a 40 percent increase over the original estimated budget of $552 million (FY2000) adopted in the Bilateral ALMA Agreement! What accounted for the increase? What were the options for recovery: reduced scope, more funding, or some combination of both? Unless a path to completion could be found, the project was dead.

The largest increases were in Management, Site Development, Antennas, and System Engineering. Two of these increases were largely the result of a failure to recognize the complexity of the tasks. Managing a large, international project turned out to be much more difficult and expensive than originally

Table 8.1

Level One WBS Task Integrated project team (IPT)	Original 2002 Estimate Cost (1,000s) $FY(2000)	11/2005 Estimate Cost (1,000s) $FY(2000)	Cost Increase (1,000s) $FY(2000)	Percent Increase (%)
Management	17,313	46,900	29,587	171
(Overhead on labor)[1]		35,670	35,670	n/a
Site development	70,049	117,682	47,633	68
Antennas[2]	227,739	319,966	92,227	40
Receivers	108,982	116,684	7,702	7
Back end	49,765	55,886	6,121	12
Correlator	14,856	10,288	−4,568	−31
Computing	34,468	37,504	3,036	9
System engineering	20,125	43,916	23,791	118
Science	9,173	9,785	612	7
Total	552,470	794,281	241,811	44

[1] Overhead on labor was distributed across the IPT categories in 2002; listed separately here.

[2] The 2002 cost was for 64 antennas; the 2005 cost is for 50. The costs in 2005 for other IPTs assume 50 antennas and are net the consequent savings.

anticipated, requiring more staff, face-to-face-meetings, and communications than a one-nation project. In addition, the decision to build the ALMA headquarters in Santiago shifted the cost of housing the JAO from operations (rent) to construction. System engineering was woefully underbudgeted for a project as complex as ALMA. Making sure all the components worked together, were delivered to specifications, and passed commissioning tests required many more people than was thought at the start. The challenges posed by the high-altitude site were simply not appreciated in the beginning. Early exploration of the site was an exhilarating lark compared to the difficulties of moving earth for roads, pouring concrete for antenna pads, laying cables, and erecting the building to house the ALMA backend and correlator. Labor at 5,000 m elevation is slow and hard, and thus expensive. These difficulties were compounded by an economic boom in Chile that drove up building costs. The cost for the antennas increased significantly due to rapidly escalating costs for steel and for oil, the raw material for carbon reinforced epoxy. The increased cost for 64 antennas over the amount budgeted in 2002 was so large as to seem unaffordable to the project management under any reasonable scenario. Accordingly, in the budget of 8 September 2005, as presented to the ALMA Board in their 15 September 2005 telecon, the number of antennas was reduced to 50 and the budgets for receivers and the backend system were adjusted accordingly. Beasley has written an account,[10] dated 26 September 2005, of the evolution of project budgets leading up to the new estimate.

Along with the new proposed budget estimate, the ALMA Board was given a menu of options for further cost savings, beyond the reduction in the number of antennas. The Board also received the report[11] of the Science IPT and ASAC containing its advice on the proposed reduction in scope. The ASAC recognized the necessity of reducing the number of antennas to 50. In choosing which other options to select, the Board followed the advice of the ASAC.[12] Savings of an additional $17 million (FY2000) were realized by reducing the number of antenna pads from 216 to 175, reducing the scope of the residence quarters at the OSF, and eliminating the antenna hanger on the high site. Budgets for furniture and travel were cut by twenty percent. Road maintenance was outsourced, moving the expense for equipment to operations. These savings were relatively small compared to what was required, but they demonstrated good faith in that everything, not just the number of antennas, was on the table for cost cutting consideration.

The next step was acquiring the additional funding implied by the new cost estimate. That required establishing the credibility of the estimate. A cost review was scheduled by the ALMA Board for October 2005 in Garmisch Partenkirchen, Germany, to be chaired by Steve Beckwith (MPIfA), who was well acquainted with the ALMA project from previous service to ESO as chair of its Science Technical Committee. It was a major event, with thirty-six participants from the three ALMA Executives twenty participants for the review panel itself, and more from NSF and the ESO Board. The review came to be known as the ALMA Cost Review (ACR). On the first day, the panel heard presentations about the project in general and then from each of the IPTs. The second day was devoted to executive sessions in which the panels drilled deeper into the costs derived by each IPT and held discussions that would lead to a report. The report[13] dated 21 November 2005 was submitted to the ALMA Board for acceptance. The report's executive summary concluded with this statement: "*In summary, the committee believes ALMA can be built to the current cost estimate, assuming resources are available, and providing that the execution of the program is robust at all levels of the project.*"

The report contained numerous recommendations, with specific recommendations for each IPT, but the bottom line was that the panel believed the cost estimate to be credible. That estimate had been prepared with the assumption that all 50 antennas would be of the same design. By late 2005, contracts had been let for the purchase of two designs, 25 each. A potential consequence of multiple designs was increased costs, both for construction and for operations. Accordingly, as was mentioned in the previous chapter, a review called the ALMA Delta Cost Review (ADCR) was held to review the project's estimate of those costs. The review was held in Balston, Virginia, near NSF headquarters, on 26 January 2006. The panel[14] was a subset of the ACR panel. They confirmed the additional funds required, completing a package that could be presented to the NSF.

One of the authors (Dickman) had played a major role as an ALMA Board member in organizing the two reviews just discussed above. Now as the person responsible for ALMA at AST/NSF, he wanted to build a bullet-proof package that asked for increased funding. That required two more elements, both independent of the ALMA Project itself. First, Wayne Van Citters, Director of AST, asked the Committee on Astronomy and Astrophysics of the National Academy of Sciences to assess the impact on ALMA of a reduction in the number of antennas.[15] Van Citters specifically asked the CAA to address four questions:

- *What would be the impact on the attainability of the level-1 science requirements?*
- *What would be the loss of speed, image quality, mosaicing ability, and point-source sensitivity? (A parametric representation of these performance changes would be welcome.)*
- *Would ALMA still be sufficiently transformational to warrant continued support by the United States?*
- *Is there a particular threshold in the number of antennas, below which ALMA would suffer a significant degradation in its performance in the above or other relevant scientific areas sufficiently serious to warrant attention?*

The CAA committee, chaired by Meg Urry of Yale University, included these conclusions in their report[16]:

- *The committee concludes that two of the three level-1 requirements, involving sensitivity and high-contrast imaging of protostellar disks, will not be met with either a 40- or a 50-antenna array. It is not clear if the third requirement, on dynamic range, can be met with a 40-antenna array even if extremely long integrations are allowed for.*
- *The committee concludes that speed, image fidelity, mosaicing ability, and point source sensitivity will all be affected if the ALMA array is descoped. The severest degradation is in image fidelity, which will be reduced by factors of 2 and 3 with descopes to 50 and 40 antennas, respectively.*
- *The committee concludes that despite not achieving the level-1 requirements, a descoped array with 50 or 40 antennas would still be capable of producing*

transformational results, particularly in advancing understanding of the youngest galaxies in the universe, how the majority of galaxies evolved, and the structure of protoplanetary disks, and would warrant continued support by the United States.

The report was in rough agreement with that of the ASAC – 50 antennas were sufficient for transformational science – but it had the *imprimatur* of the National Academy of Sciences and would carry much more weight at NSF in an argument for approval of the rebaselined budget with its requirement for a forty per cent increase in funding.

Next, NSF commissioned yet another cost review, this one reporting to NSF rather than to the ALMA Board. The review was called the NSF Cost/Management Review (NCMR). It was chaired by Donald Hartill of Cornell University and was held at NRAO headquarters in Charlottesville on 30 January–1 February 2006. The NCMR panel was provided with the reports of the ACR and ADCR as background material. It also heard presentations from project management, each IPT, and of the operations plan. At the close of the review, the panel made twenty observations,[17] the first 10 of which they judged already to be under control and 10 more that required close attention: increasing the contingency, safety on the high elevation work sites in Chile, and management recommendations in a number of areas. But most importantly, the panel validated the cost estimates of the ACR and ACDR. The panel opined that no more than $14.2 million of the increase was due to project failure, namely, the failure to complete prototype antenna testing in a timely manner, which caused a delay in the antenna procurement.

Additional budget relief arrived in December 2005. An MOU[18] was signed, whereby, ASIAA of Taiwan would join the United States and Canada in the North American partner to ALMA, while they also continued to collaborate with Japan. Taiwan would contribute the cost of two to three fully equipped antennas ($30 million current year dollars over 10 years) and make an appropriate contribution to operations. A goal was to deliver a minimum of $15M in the form of equipment made in Taiwan. In return, Taiwanese astronomers could compete for the North American share of the observing time and participate in ALMA committees. This initiative was driven by Mike Turner, following his strong belief in ALMA as a facility capable of transformational science. Although the amount of money was not large compared to the total project cost, acquiring a new ALMA partner for North America sold well in the NSF Director's office.

Dickman had informed the ALMA Board at its 27 January 2006 meeting of the necessity of the Hartill Review if NSF was to approve the cost increase.

Catherine Cesarsky expressed concern that the process would delay ALMA construction. The Board was assured that the results of the review would be presented to the NSF Director in March 2006, and, pending his approval, to the National Science Board the following April, in time for inclusion in the NSF budget request for FY2008. A plan would be made for keeping the project on schedule. Dickman briefed[19] NSF Director Arden Bement on 2 March 2006 and the meeting was a success. On 7 April 2006, Bement sent a memorandum[20] to the NSB recommending approval of the new ALMA budget. It met with NSB approval, to the gratification of Dickman who had orchestrated much of the process. The NSF budget request to Congress for FY2008 asked for the necessary funds. They were appropriated by the US Congress as a routine matter. Furthermore, the request had outlined the requirements for the outlying years, so subsequent requests would also be routine, following the funding plan.

The resolution of the rebaselining crisis revealed a stark cultural difference between the ALMA parties. ESO's guaranteed annual income allowed it to cover the cost overrun by extending the schedule of payments by two years. This required Council approval, which was granted after consideration of the report of the ACR and confirmation that ALMA was affordable. By comparison, the US process was fraught with uncertainty. Its eventual success was due to a carefully prepared and validated justification presented to top NSF management. Both ESO's and NSF's steadfast commitments to advancing the frontiers of science through large, multiuser facilities carried the day. Japan's commitment was never in doubt. A major crisis had passed. With the funding for ALMA as secure as it could ever be, the project could continue construction uninterrupted. All that remained was to continue putting ALMA together and then make it work.

Notes

1 Wootten, H.A. 2005, ALMA Gains Capabilities with Japan's Entry, NRAO Newsletter #102 http://library.nrao.edu/public/pubs/news/NRAO_NEWS_102.pdf.
2 Clancy to Vanden Bout, private communication.
3 Michael Ledford of Lewis-Burke Associates to Paul Vanden Bout, private communication.
4 From the report of the Senate Appropriations Committee, Departments of Veterans Affairs and Housing and Urban Development, and Independent Agencies Appropriations Bill for FY2001, B. Mikulski, 20 July 2000. www.nsf.gov/about/budget/fy2001/pdf/tools.pdf.
5 McGuire to Vanden Bout, private communication. Carol McGuire was one of Senator Domenici's aides on the Senate Budget Committee staff.

6 NRAO did do work for ALMA at its Socorro, New Mexico, site, building the ALMA "backend," the system that digitizes the signals from the antennas and sends them to the correlator, and hosting the group that wrote the software for the ALMA real-time operating system that controls the array during observations.

7 The list and copies of many of the letters can be found at NAA-NRAO, ALMA, ALMA Planning, Box 2.

8 From the report of the Senate Appropriations Committee, Departments of Veterans Affairs and Housing and Urban Development, and Independent Agencies Appropriations Bill for FY2002, B. Mikulski, 20 July 2001. www.nsf.gov/about/budget/fy2002/pdf/tools.pdf.

9 The FY(2002) funds for ALMA construction were added to NSF's MRE account. The Major Research Equipment (MRE) account became the Major Research Equipment and Facilities Construction (MREFC) account in the FY(2003) NSF budget, the source of all succeeding US funds for ALMA construction.

10 The account by Beasley of the ALMA cost increase evolution can be found at NAA-PVB, ALMA, ALMA: The Story of a Science Mega-Project. https://science.nrao.edu/about/publications/alma.

11 The report can be found at NAA-PVB, ALMA, ALMA: The Story of a Science Mega-Project. https://science.nrao.edu/about/publications/alma.

12 A spreadsheet with all the cost savings options that shows those the ALMA Board approved can be found at NAA-PVB, ALMA, ALMA: The Story of a Science Mega-Project. https://science.nrao.edu/about/publications/alma.

13 The report *Review of Costs for the ALMA* (Beckwith Report) can be found at NAA-NRAO, ALMA, ALMA Rebaselining. https://science.nrao.edu/about/publications/alma.

14 The panel members included: Steve Beckwith (MPIfA), Jim Crocker (Lockheed-Martin), Thijs de Graauw (ESTEC), Peter Dewdney (DRAO), Tom Phillips (Caltech), and Jean Turner (UCLA).

15 Van Citters' letter is Appendix A of the CAA report.

16 The CAA report, *The Atacama Large Millimeter Array – Implications of a Potential Descope*, can be found at https://nap.nationalacademies.org/download/11326.

17 The NCMR report, known as the "Hartill Report" may be found at NAA-PVB, ALMA, ALMA: The Story of a Science Mega-Project. https://science.nrao.edu/about/publications/alma.

18 The MOU (unsigned copy) for the Taiwan-American Program in Radio Astronomy (TAPRA) can be found at NAA-NRAO, ALMA, ALMA Multi-Institutional Agreements. https://science.nrao.edu/about/publications/alma.

19 Dickman's presentation can be found at NAA-PVB, ALMA, ALMA: The Story of a Science Mega-Project. https://science.nrao.edu/about/publications/alma.

20 Bement's memorandum can be found at NAA-PVB, ALMA, ALMA: The Story of a Science Mega-Project. https://science.nrao.edu/about/publications/alma.

9

Construction and Inauguration

"Keizoku wa chikara nari (Persistence pays off)"

Japanese proverb

No telescope of the scale of ALMA had ever been built on a site as high as 5,000 m above sea level. Working effectively in the thin atmosphere with only 50 percent of the oxygen available at sea level would require special equipment and procedures. Furthermore, the site was remote, lacking even the most basic necessities such as roads, water, and electricity. The components of ALMA[1] were manufactured in far-flung facilities in Japan, North America, and Europe. As they arrived at the site they needed to be inspected, installed, tested, and transported as outfitted antennas to final locations. Much of the work was repetitive. Fifty-four 12 m plus twelve 7 m antennas all received the same set of electronics. It was a massive task that required determination and focus, despite the repetition. It all had to happen by the inauguration, scheduled for 2013.

Groundbreaking

The ALMA groundbreaking ceremony was held on 6 November 2003, attended by representatives from all the partners. After the observance of a Pachamama ("Mother Earth") ground blessing ritual conducted by a local shaman and a prayer by Padre Ibar Astudillo Godoy, parish priest in the Cathedral of Antofagasta, the group shown in Figure 9.1 posed for a photograph to record the event. Riccardo Giacconi, President of AUI, gave a speech[2] that highlighted the major observatories of modern astronomy, on the ground and in space. He ended his remarks by thanking the host country: "*I would like to end by thanking our host, Chile, for joining us in this noble voyage of discovery and making this enchanted land the home for some of the most important of these great enterprises.*"

Figure 9.1 Some of the groundbreaking representatives. Left to right: Bob Dickman, Eduardo Hardy, Fred Lo, Massimo Tarenghi, Catherine Cesarsky, and Daniel Hofstadt. On the concrete block in the foreground are the remains of the Pachamama ceremony. Courtesy of Ineke Dickman, reproduced by permission.

Basics

Roads – In the early days of exploration and testing, the ALMA site was reached by one of three routes. One could drive up the road to the Jama Pass, Route 27, and either take the cutoff to Cerro Toco and then down to the site, or drive further, around Cerro Toco and Cerro Chajnantor, to Pampa la Bola, where the Japanese would install two submm telescopes, and thence back west to the site. The third option was to drive directly east from Route 23, starting at a point between San Pedro and Tocanao, straight up to the site on a mining road that led to an abandoned sulfur mine on Cerro Negro. Which of these routes would ALMA use to transport antennas from their assembly point to the high site? Using state highways would save the cost of road construction. But it would require permits and police convoys to ensure safety as the transporters, moving at 10 km/h, would tie up traffic. This option was judged to be unworkable. Instead, a road was built roughly along the track to Cerro Negro from Route 23 to a site at 2,900 m elevation where the antennas were to be assembled and from there to the Array Operations Site (AOS) at 5,000 m. The Operations Support Facility (OSF), would be the location of lodging for contractors and

ALMA staff, cafeterias, a building for the assembly of the North American antennas, and the large OSF technical building, which would house offices, electronics laboratories, and an array control center. At the AOS, 175 antenna pads needed to be built and connected by roads. Road construction began in 2004 and was completed and accepted in March 2013.

Electrical Power – The Chilean national power grid does not reach San Pedro, so the possibility of connecting to a source of power in the village was not possible. Initially, diesel power generators were leased and installed at the OSF. The diesel fuel came from a refinery in Concepción, over 2,000 km south of San Pedro. It was rumored that during construction there were five fuel trucks on the road going to or from ALMA at any given time. The plan was eventually to connect ALMA to the national grid by constructing towers and high voltage lines to carry the power from Calama to San Pedro and then to ALMA. The cost for this turned out to be far more than had been budgeted. An alternative was to purchase generators to replace the growing set of rental units. This occurred in 2011, quite late in the construction. ALMA now runs on a 5.7 MW Taurus 60 turbine generator by Solar Turbines, a Caterpillar company, that burns liquified petroleum gas (LPG), but can be converted to burn diesel fuel. A standby unit is ready to take over in case of a failure. And a third unit is available to be deployed as a standby when one of the units is undergoing maintenance. At the 2,900 m elevation of the OSF one of these units can only generate 3.75 MW of power. The power plant is shown in Figure 9.2. Electrical power is a significant item in the ALMA operations budget. Sixty percent of the power is used to run the antennas, mainly the refrigerators that cool the receiver electronics to 4 K, with the balance used by the OSF. A crew of nine people is required to operate the power station and maintain the 27 kV transmission grid, daily, around the clock.

Security – It is impossible to fence off the huge area on which ALMA sits. But access other than through the main entrance is limited by the natural terrain to all but the most adventuresome. The entrance to the OSF from Route 23 is secured by a guard house near the highway. It is still possible to access the high site using the mining road that goes along Cerro Toco from Route 27 (the road used to make the first site search visits). Also, one can access the site from the Japanese telescopes at Pampa la Bola behind Cerro Chajnantor. And, of course, one can simply drive cross-country in a four-wheel drive vehicle, working one's way to the high site, as was done in the early days of the site exploration. But unwanted visitors can be spotted by a network of 16 security cameras or by ALMA staff working on the site. Even so, the high site is enormous and there have been a limited number of incidents of vandalism.

Figure 9.2 The ALMA power plant built by TSK Elektronica y Electricidad S.A. The plant uses Taurus 60 gas turbine generators, by Solar Turbines. The three 60,000 L fuel tanks on the right side of the photograph store diesel fuel; an additional three tanks, not visible, store 200,000 L each of LPG, enough to run ALMA for three days. Credit: ALMA/ESO/AUI/NINS, CC BY 4.0.

Safety – The ALMA safety program has its roots in the early site search for the MMA. As the search progressed toward higher and higher sites, the question of safety at high elevations arose. Dr. John West, an expert on high-altitude medicine, was hired as a consultant. In 2006, the NRAO safety officer, Jody Bolyard, spent a year in Chile writing the ALMA safety program. It was based on his experience with the MMA search and Dr. West's advice. There are three main functions of the safety program: training, inspections, and accident mitigation. The incidence of accidents is limited by rigorous training, alcohol restrictions on site, and the establishment of protocols for dangerous procedures.

The two unfortunate accidental deaths that occurred during ALMA construction were both the result of drivers losing control of their vehicles. Alcohol is restricted at ALMA, and drivers are given a breathalyzer test on entering the site. Another safety issue is fire, which is difficult to fight at high elevations due to the limited availability of water and the stressful conditions for the fire fighters in the thinner air. ALMA has trained volunteer firefighters who are well equipped to deal with blazes that might occur. To mitigate the effects of high elevation, workers at the high site use supplemental oxygen, as do the transporter drivers. All visitors to the high site carry oxygen canisters, either in a specially equipped backpack or handheld. The air in the building that houses the correlators is enhanced in oxygen content. In case of emergency, so-called Gamow Bags are available should someone suffer high-altitude pulmonary or cranial edema. Victims can be sealed in the bags which are then inflated to sea-level pressure and transported to a hospital. Fortunately, they have never been used.

An on-site ALMA rescue squad with paramedics is trained to handle accident victims. Victims can quickly be treated in the ALMA Polyclinic, which has offices on-site at the OSF technical building, residencia, and at the AOS. The rescue squad is also ready to help with accidents at the other astronomical facilities in the Science Reserve as part of a good neighbor policy. The record of accidents at the ALMA site is gratifyingly low, given the hazards and the number of ALMA staff and contractors that inhabit the workplace. Since safety is such an important concern, trainings are regularly offered to ALMA staff on topics such as high-altitude first aid, fire and emergency preparedness, and off-road driving.

Water – Although the Atacama Desert is a very arid region, it does rain. Total precipitation on the ALMA high site, largely in snow, is equivalent to 10–30 cm of water annually. There is an abundance of ground water at lower elevations that has flowed down from snow melt at higher elevations. This water is close to the surface in the oases like San Pedro and other villages at the base of the Andes. It is used for irrigation but is not recommended for human sustained consumption due to its arsenic content. It is thought many of the health problems of the indigenous Atacamaño population result from their consumption of arsenic-laced water. ALMA buys water from San Pedro, transported from the village by truck to the OSF. ALMA uses about 75,000 L of water per day. This may seem like a lot of water until one realizes that it must supply: a hotel with 120 guests expecting to eat meals, shower, and enjoy clean linens; the central OSF building and other facilities; and for cleaning antennas, transporters, and other equipment of accumulated dust. An additional 100,000 bottles of drinking water are purchased annually. Waste water is processed in a treatment plant. The processed gray water is used to keep down dust on the roads and construction sites. Excess gray water is safely discharged into the environment.

Communications – A mechanism for transferring data from the high site to the world at large was needed from the earliest days of site exploration and study. Initially, the site study data were simply downloaded to a storage device, magnetic tape at first and then compact disks. These were retrieved every few months by Angel Otárola and express shipped to Simon Radford for analysis. This brute force method was supplemented by an Inmarsat satellite phone link which allowed for daily data downloads. During ALMA construction, a microwave link to Calama was established, where a connection to the internet was available.

For the massive amounts of data from ALMA, an optic fiber connection was required, connecting ALMA to San Pedro and thence to Calama and Antofagasta. The fiber connection was installed by Silica Networks in 2014 as part of the first project of the ALMA Development Program, the ongoing program whereby

ALMA is improved and enhanced with new equipment. Silica Networks was unable to charge for ALMA's use of the fiber until it received government approval, a process that took many years while ALMA enjoyed free service.

The Silica link allowed ALMA to connect to the existing data system through northern Chile that came from a 2010 European investment known as EVALSO (Enabling Virtual Access to Latin American Southern Observatories), which was made in partnership with Red Universitaria Nacional (REUNA) in Chile. EVALSO was designed to enable faster connection for ESO's facilities at Cerro Paranal and Armazones, where the ELT would be constructed. In 2012, ALMA signed an MOU with REUNA to transport data from Antofagasta to the JAO in Santiago, where it is redirected to the various regional centers. Beyond the Santiago metropolitan region, it is the responsibility of the ALMA partners to transport the data to the three ALMA Regional Centers (ARCs), where copies of the data are stored in addition to the archive at the JAO. ALMA data are sent to Europe via ESO's established network. In the case of North America, data travel via Brazil to the Florida International University[3] (FIU) in Miami where they connect to the US National Research and Education Network backbone for further transport to the North American ARC in Charlottesville. The data take a longer journey to the East Asian ARC, following the same path as the North American data to FIU and connecting there via the fiber optic carrier Internet2 to Tokyo.

High Site

Antenna Pads – The fifty-four 12 m antennas can be moved into numerous distinct configurations by utilizing the 175 antenna pads. There are an additional 16 pads for the Japanese compact array and four 12 m antennas. The pads incorporate kinematic mounts whereby the placement of an antenna on the pad always results in the antenna sitting in the exact same location. Clamps tie the antennas to their pads to secure them against high winds. Power is supplied to the antenna through cables that connect the pads to the generators at the OSF. The signal received by an antenna during observing is transmitted to the OSF via optic fiber cables running to each pad. The pad connections required a significant amount of trenching, a challenge in the abrasive, volcanic earth. As an example of what can go wrong, a simple purchasing error resulted in the connections from power transformers to 51 antenna pads to be in noncompliance with the grounding standard specified by the antenna manufacturers. Rewiring these pads was necessary to maintain the antenna warranty. It was impossible to dig up the faulty power cables without damaging the fiber optic cable. The solution was to bury an additional ground wire as deep as possible. This was done hydraulically, using heated water to cut a trench.

Figure 9.3 ALMA antenna transporter *Otto* carrying the last antenna, built by AEM, to the high site to join the array. Credit: ESO, CC BY 4.0.

Antenna Transporters – Being able to relocate antennas to different pads and generate distinct configurations, some with vast separations between antennas, is a vital function for interferometry. It makes possible observations with higher angular resolution, that is, greater detail in the images obtained. Moving the antennas closer together provides greater sensitivity to extended emission from a source. ALMA does not have fixed configurations. The 12 m antennas can be spaced as needed for the scientific goals of projects, from a minimum separation of 15 m to a maximum of 16 km, in what is known as a dynamic array that is changing, sometimes daily, throughout an observing cycle. Each 12 m antenna weighs approximately 100,000 kg, so moving one about requires a very special transporter; ALMA has two such custom-designed transporters, shown in Figure 9.3. Each is 20 m long by 10 m wide by 6 m tall, weighs 260,000 kg, and has 28 independently steerable tires. To learn to drive a transporter one must take a six-month course, and only a few critical employees are so equipped. The drivers use supplemental oxygen while driving to stay alert at elevation. Each transporter has two 500 kW diesel engines that only deliver 320 kW in the thin air at 5,000 m elevation. The engines are supplied fuel from two 1500 L tanks. Electrical power is supplied to an antenna from the transporter during a move. This keeps the refrigerators running that cool the receiver electronics and allows the antenna base to be driven to the proper orientation when it is being placed on a pad.[4] The transporters live in a hangar at the OSF when they are not in use. A second shed on the high site is available for overnight shelter if necessary. The transporters are named Otto and Lore, the given names of the Scheuerle's, who own the company that made the transporters, Scheuerle Fahrzeugfabrik GmbH, of Pfedelbach, Germany. In December 2007, the transporters were loaded onto a barge to be shipped down rivers and canals to Antwerp, Belgium, and then by ship to Antofagasta, Chile. They were then unloaded and driven to the ALMA site in a convoy that occupied the city street and highway lanes in both directions. The trip was made possible with the assistance of the Carabineros.

Buildings

Operations Support Facility – The OSF is the complex of facilities required to operate ALMA. It is located in a 1 km^2 area owned by the ALMA partners at an elevation of 2,900 m, midway between the valley floor and the high site. The main facility is the 7,000 m^2 Technical Building, shown in Figure 9.4, that contains offices, laboratories, the array control center, and emergency services. It dominates the OSF in a central location. The architect, Fichtner GmbH & Co, Germany, gave the building a sleek, modern design that blends into the landscape. The construction contractor was the consortium Vial y Vives and Mena y Ovalle, Ltda, of Santiago, Chile. Design work began in November 2003, construction on 10 August 2006. The building was completed in 2008, but outfitting continued for two more years as activities ramped up.

Figure 9.4 An aerial view of the OSF. Credit: ESO, CC BY 4.0.

Figure 9.5 The AOS Technical Building. The arches in the foreground are vehicle shelters. The large shed in the background to the right is the transporter shelter. Credit: ESO, CC BY 4.0.

AOS Technical Building – The supercomputers, known as correlators, that combine the signals from the array antennas to form an image are housed in the main building at the high site, one for the large array of 12 m antennas and one for the compact array of 7 m antennas. The technical building also houses the electronics that synchronize the antenna receivers in time, the so-called "local oscillator" system. The 5,000 m elevation imposes special requirements. Heat generated by correlators and associated electronics must be dissipated by a refrigeration system designed to operate in thin air. The air in the building is charged with supplemental oxygen to allow technicians to function normally in servicing the equipment. The building is the highest in the world with this level of technical activity. An adjacent shed provides cover for transporters should they need to stay at the AOS. The building was designed by the international architectural firm M3 and built by Constuctora Tesca Con.Pax S.A. of Santiago, Chile. It is shown in Figure 9.5.

Residence Hall – During construction, ALMA staff and contractor employees were housed in separate, temporary residences, each with its own cafeteria. On 23 February 2015, ground was broken for the ALMA Residencia. In Massimo Tarenghi's dream for this "hotel," it was to bridge a ravine, with a swimming pool built into the ravine below. His hope was to build something at least as stunning as the Residencia at ESO's Paranal Observatory. Sadly, there were

Figure 9.6 The ALMA residence hall provides living accommodations for up to 120 staff members and visitors along with a cafeteria, and a variety of venues for exercise and relaxation. Credit: ESO, CC BY 4.0.

insufficient funds to realize this vision. The more practical residencia, shown in Figure 9.6, that was built has an excellent design that was created by the Finnish architectural firm Kuovo and Partanen. The Chilean firm Rigotti and Simonovic, Arquitectos, adapted the design to local standards. The residence hall was built by the consortium AXIS LyD Construcciones Ltda, consisting of Constructora LyD S.A. and Axis Desarrollos Constructivos S.A., two Chilean companies that had extensive experience in constructing residential buildings in the challenging environment. The residence hall can house up to 120 employees and visitors. OSF employees work on the *sistema de turno*,[5] a well-established program in Chile of rotating shifts whereby they are at the OSF for five days and then are at home for two days, or for positions that require continuous duty, at the OSF for eight days and home for six days. Many employees commute to the OSF from long distances, as far as Santiago and beyond. ALMA provides for their transportation from specified pick-up points and for their room and board. The residencia provides amenities for its guests that help maintain physical and mental health outside of work hours: a swimming pool, gymnasium, and sauna.

Antenna Assembly Building – The contractor for the North American antennas, General Dynamics Satcom Technologies (Vertex), wished to assemble them in a controlled environment, out of the weather. Accordingly, a large building, shown in Figure 9.7, was erected at the southwest corner of the OSF for that purpose. Four antennas could be assembled simultaneously in the building, the work aided by lift buckets and an overhead gantry. An additional three antenna pads are located outside the building to store finished antennas. The downhill side of the building has huge doors, large enough to allow a transporter loaded with an antenna to pass through. The building has proven to be extremely useful in ALMA operations for antenna and transporter maintenance, justifying the decision not to remove it once the North American antennas had been assembled. It has, however, been an item of controversy with the local residents, particularly, those living in the village of Toconao, directly below the site. The source of the controversy is an illustration of unexpected consequences. When first constructed the building did not have doors; it was open, facing Toconao. The temporary electric power generators in use at that time had to be run with a power load. Running without a load caused breakdowns. So the lights on the OSF site were left on all night, whether or not they were needed, including those in the antenna assembly building. This spoiled the night view of the mountains above Toconao, the tranquil scene the residents were accustomed to seeing replaced by one dominated by the maw of a harshly illuminated building interior. The eventual installation of the doors helped considerably.

Joint ALMA Office – As was described in Chapter 7, the ALMA Observatory offices are housed in a building on the ESO campus in Vitacura, Santiago. The building provides office space for administrative and scientific/technical staff, meeting rooms, and an auditorium. An underground garage can hold 150 cars, shared with ESO next door. The location is secure and in a pleasant neighborhood,

Figure 9.7 The Vertex antenna assembly building is on the left. Three North American antennas can be seen stored on the outdoor pads. The MELCO assembly area is in the center, where four of the antennas for the Atacama Compact Array (ACA) can be seen. The AEM assembly area for the European antennas is on the right, where seven antennas appear to be complete and more reflecting surfaces are being assembled. Credit: ESO, CC BY 4.0.

close to the Vitacura Municipal Center and the expansive Parque Bicentenario nestled along the Mapocho River. Activity at the JAO has steadily increased since operations began. For example, the arrays can now be operated from a control room extension there.

Antennas

ALMA's 66 antennas are its most expensive, mechanically complex, and visible components. They were fabricated in distant facilities around the world, far from Chile, by three different companies in four different designs to exacting specifications. Formal quality assurance (QA) procedures were established to ensure that each antenna with all its sub-components would actually work as specified. These procedures also applied to the signal correlators and their associated electronic systems, and to site construction of roads, buildings, and utilities. The larger the project, the greater the importance of QA. ALMA never had enough. As is typical for a large project, the ALMA quality assurance procedures had names and acronyms.

Preliminary Acceptance In-house – To ensure that nothing faulty was shipped from its place of fabrication, every deliverable from an IPT was tested by a PAI team prior to being sent to the ALMA site.

Provisional Acceptance On-site – The PAS team was the component of the JAO responsible for accepting the deliverables by checking that nothing had been damaged during shipment, before assembly and integration into working telescopes. The PAS team was also responsible for inspecting and accepting the construction work on the ALMA site. The JAO was severely understaffed at the beginning of construction. Pressure to complete the integration of electronics and make the antennas ready for observing meant that provisional acceptances were issued by the hard-pressed PAS group, reserving the right to make a final inspection and acceptance later. For example, the roads on the site had been in use for years before a final inspection occurred. This led to a conflict with the contractor over who was responsible for bringing the roads up to specifications. It was only one example of contract and warranty disputes that cost the ALMA Executives a great deal of time and effort to resolve. The contractor that constructed the outdoor antenna pads at the OSF provided another example. To allow for water to run off, they were to have a two percent slope from the center to the pad edge, that is, a slope of 2:100. For the first of these pads, the contractor misread this specification in the drawings as 2 degrees, a slope that is nearly twice as large. The first time an antenna was placed on one of these pads, it teetered on the tip of the concrete cone.

Assembly, Integration, and Verification – The AIV team was the component of the JAO responsible for assembling a functional antenna, that is, equipping an antenna with the electronics systems that made it into a telescope, making sure the systems worked together, and confirming that their functionality met specifications. The AIV team had to work closely with IPTs responsible for fabricating and delivering the systems to fix problems and work out bugs.

Commissioning and Science Verification – The CSV team of the JAO was responsible for making test astronomical observations with a completed antenna, verifying that it was ready for service in the array. The CSV team was strongly supported by the AIV team. Initial observations with a new telescope, so-called "first light" observations, often reveal unexpected results that prompt a variety of tests that may lead to still others. The activity can become a somewhat interrupt-driven seat-of-the-pants exercise with some appearance of disorganization. If this was true of the CSV team, perhaps it was so because the tasks were conducted by astronomers, who typically are less management oriented than engineers, and the team was battling against time in order to get to the real science as quickly as possible. Throughout, the CSV team also tested new capabilities and investigated any issues that arose with data acquisition. What counts is that ALMA works as specified, for which the team, led by Richard Hills, deserves immense credit. Following Cycle 0, and as ALMA approached full operations, the CSV team successfully morphed into a team known as Extension and Optimization of Capabilities (EOC).

Vertex Antenna Assembly – The following description of this process focuses on the North American antennas, for which the authors had access to the most extensive documentation. The North American antenna components and materials are shown in Figure 9.8. Vertex was the contractor for the antennas and these components came from its different facilities and subcontractors: the Invar[6] cone, cabin, azimuth structure, and pedestal were shipped from the Vertex plant in Kilgore, Texas, after being fabricated in Mexia, Texas; the CFRP components – Backup Structure (BUS) and quadrapod support legs – from Airborne in The Hague, The Netherlands; the reflecting panels from Zrinski AG in Wurmlingen, Germany; and the machined aluminum subreflector from CPI Vertex Antennentechnik (VA) in Duisburg, Germany.

The delivery schedule for all the antennas was specified in their contracts. The first of 25 Vertex antennas arrived at the OSF on 24 April 2007 and the last was completed and accepted on 16 November 2012. The first of 25 AEM antennas arrived about a year later in 2008. The AEM delivery schedule was faster paced to make the 2012 goal of having all the 12 m antennas accepted. With only 16 antennas, MELCO enjoyed a somewhat more relaxed schedule for delivery, easily meeting the 2012 deadline. Re-assembly and testing of the antennas was a learning process. For example, the first Vertex antenna took about 434 days to reach acceptance. After that, the time required dropped steadily to 184 days for the last antenna. It should be noted that several antennas could be in assembly at the same time. The Vertex building had four assembly pads; AEM and MELCO had multiple assembly pads as well.

The activity prior to shipment for the Vertex antennas was located in two places. In a Vertex facility in Kilgore, Texas, the mild steel components and INVAR cone were fitted together. Before attaching the azimuth structure to the pedestal, cabinets for the main power distribution system and uninterruptible power supplies were installed in the pedestal, along with the azimuth cable wrap, a mechanism to prevent cables going to the receiver cabin from binding when the antenna moved in azimuth. Air-conditioning equipment was installed in the receiver cabin prior to attachment of the INVAR cone. Two access platforms were assembled and test fitted to the assembly. Then the required cabling and piping were installed. All exposed steel and INVAR surfaces were covered with closed-cell foam insulation and then a layer of aluminum cladding. Finally, the drive system was tested. While the North American Antenna IPT monitored the assembly and testing particularly closely for the first antennas, the process was truly a world-wide affair.

After completion of factory acceptance testing, the assembly was placed horizontally on a heavy steel pallet or skid and wrapped for shipping. The package came to be called the "Blob." Its size and weight were comparable to, if

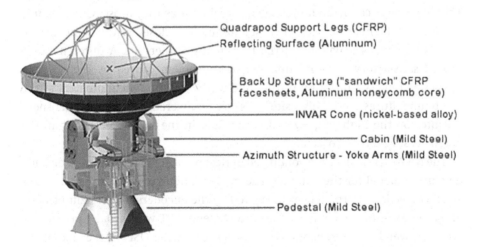

Figure 9.8 Diagram of the subassemblies of the North American antenna indicating the materials from which they were fabricated (CFRP is the acronym for carbon fiber reinforced plastic). Credit: Art Symmes; NRAO/AUI/NSF, CC BY 3.0.

somewhat less than, those of a diesel locomotive. The load traveled on a high-capacity flatbed trailer to the port of Houston, Texas, where it was loaded onto a cargo ship for shipment to Antofagasta, Chile. After being placed on a different skid and again on a flatbed trailer, the Blob headed to the OSF, filling all lanes of the city streets and highways along the way. As with the antenna transporters, the Carabineros provided traffic control and security for the convoy. Two 20 ft shipping containers with the access platforms and air conditioning equipment accompanied the Blob. The entire trip from Kilgore to the OSF took about four weeks.

All the antenna components above the INVAR cone were assembled in Duisburg, Germany, at VA. The BUS of an antenna is made up of plates that consist of aluminum honeycomb sandwiched between CFRP sheets. The aluminum reflecting panels are mounted to this structure. The rigidity of the structure was tested by applying specified bending loads and measuring the deflection. After testing, the structure was disassembled and packed for shipment. All the pieces fit together with location pins and were labeled with barcodes to ensure an exact duplicate re-assembly in Chile. Shipment required two 40 ft shipping containers for the BUS panels and three 20 ft shipping containers for the reflecting panels, support legs, and subreflector. The trip to Antofagasta required six weeks.

When the first Vertex antenna arrived at the OSF in 2007, the assembly building was not finished yet, so assembly began on an outdoor pad. Subsequent antennas were assembled indoors. Assembly was the responsibility of the contractor, with the NA Antenna IPT monitoring the workmanship and testing. The two groups cooperated in resolving problems as they arose. One example of such a problem arose with the first antenna. The closed-cell foam insulation installed in Texas expanded in the thin air by 50 percent. A different foam proved satisfactory. Another unanticipated problem caused by the environment was the crystallization of the fluid in the antenna inclinometers at the low temperatures on the high site. This was discovered after an antenna spent a winter on the high site. Switching the fluid in the inclinometers of all the antennas solved the problem.

The NA Antenna IPT then conducted optical pointing observations to develop a pointing model for the antenna. The model, unique to each antenna, makes small corrections to the angles measured by the encoders on the azimuth and elevation axes. In the course of this, the Antenna IPT was initially puzzled by an effect well-known to geophysicists – vertical at the OSF on the side of the mountain is not the same as on the high site on top of the mountain, but is tipped slightly, the downward end pointing a little towards the mountain. They had not observed the "deflection of the vertical" effect at the prototype test facility which was located on a broad, flat plain. And the effect is not present on the Llano de Chajnantor where the AOS is located. This exercise was the first opportunity to use the antenna drive system and numerous bugs were found and corrected by the contractor.

The contractor had set the reflecting surface to an accuracy of 35 μm using photogrammetry, a technique that involves taking pictures of the surface from several angles to obtain a three-dimensional image. The antennas then received extensive testing of mechanical alignments, servo system performance, and general satisfactory compliance with specifications. The Antenna IPT made holographic observations to set the reflecting surface to an accuracy of 20 μm. Holography is a technique that involves observing a fixed transmitter at a large variety of angles to obtain an error map of the reflecting surface.

The NA Antenna IPT served as an interface between the contractor and the AIV group, supporting the AIV group in solving little problems and bringing in the contractor for warranty issues. Nick Emerson from the IPT and Lutz Stenvers from VA conducted operation and maintenance training sessions for the AIV group. Figure 9.9 shows an assembled antenna, ready for its electronics.

Figure 9.9 An antenna, ready for its electronics, being loaded onto a transporter. It is shown in the antenna assembly building, built by Vertex for assembly of its antennas. Credit: Art Symmes; NRAO/AUI/NSF, CC BY 3.0.

Memories of Roles in ALMA/NAOJ

NAOJ formally joined ALMA two years late, and at that time the specifications for the interfaces between ALMA antennas and other subsystems were advanced. When I participated in the ALMA transporter meeting in 2004 September, I asked to modify the antenna clearance with the ALMA transporter. At first, ALMA Director Massimo Tarenghi refused the request. I explained that the interface change did not affect EU or NA antenna design and the request was eventually approved. After that, Stefano Stanghellini, Jeff Zivick, and I, with ALMA Project Manager Tony Beasley, enjoyed improved communications on antenna interface and specification issues. Things went better after that.

NAOJ ALMA project overcame several challenges by the work of several individuals: Junji Inatani led the antenna engineering discussion with MELCO on details to fulfill the stringent specifications. NAOJ ALMA project manager, Satoru Iguchi was responsible for the schedule and costs and argued with MELCO. I led the verification team with Kouichiro Nakanishi to make the verification plan and procure measurement instruments, including the holography system.

These efforts saw the first three ACA 12-m antennas arrive at OSF in 2007 September on schedule. In the initial holography measurement, the thermal deformation of the dish was larger than predicted. After examining numerous holography maps with temperature data, NAOJ identified the cause. MELCO improved the insulation of the receiver cabin and yoke structures and NAOJ finally demonstrated that the ACA 12-m antenna met the ALMA surface specification. We conducted verification measurements on pointing, surface, path length stability, and servo performance for months. The antenna was accepted as the first ALMA antenna in the acceptance review of December 2008 and moved to the JAO OSF pad the following month. The NAOJ ALMA project were very happy over this accomplishment and I think this was due to the individual strengths of the team members of the NAOJ ALMA project even under difficult circumstances.

I served as the chair of the ACA 7-m antenna Preliminary Design Review in September 2008. The design adopted a steel BUS with fans stabilizing the BUS temperature. The reviewers were initially concerned about such an unusual approach to meet stringent ALMA surface specification of 20 μm. The review panel finally accepted that design since MELCO showed the detailed supporting analysis. In closing remarks, I said that the ACA

7-m antenna would make history. Later, the holography measurements demonstrated that the 7-m antenna had the best surface accuracy of better than 5 μm. Further, the antenna achieved very small thermal deformation despite of relatively large temperature variations.

I was fortunate to have a unique and exciting experience in an extremely large international project. I appreciate my colleagues, especially Koh-Ichiro Morita who passed away just after the ALMA early science started.

Masao Saito
National Astronomical Observatory of Japan
Tokyo, Japan

The process just described was in broad outline similar to the European and Japanese antennas, and the three efforts, somewhat incredibly, converged on time to form the final array at the Chajnantor Plateau.

Electronics

An optical telescope forms an image using mirrors and lenses; a radio interferometer via electronics. ALMA's electronics fall into four major systems: receivers, local oscillator, intermediate frequency, and correlator. Together, ALMA's electronics systems represent a multitude of parts: wires, cables, integrated circuits, resistors, capacitors, inductors, micro-processors, fiber optics, power supplies, digitizers, oscillators, lasers, connecters, refrigerators, and cooling equipment, that were built into systems in laboratories around the world to be fitted onto the antennas in Chile. Electronic systems were the third largest fraction of the ALMA budget, essentially tied with the cost of developing the site at 15 percent of the total cost.

Receivers – The signals received by radio telescopes are extraordinarily weak, and ALMA enjoys no exception in that regard. The signals they detect are measured in Janskys (Jy), a unit[7] named after Karl Jansky, who was the first to discover radio waves from a non-terrestrial source. A cell phone on the Moon transmitting 1/4 W has a signal strength of 1 Jy. Modern radio telescopes routinely detect signals that are only a micro-Jansky (μJy) in strength, 10 million times weaker. For ALMA to do so at millimeter and submillimeter frequencies requires receivers that define the state of the art. The key components that make this possible are superconductor-insulator-superconductor

(SIS) junctions and high electron mobility transistors (HEMT). At NRAO, the SIS junctions and their mixers were designed by Tony Kerr and Shing-Kuo Pan, and the junctions were fabricated on silicon wafers under a cooperative arrangement with a laboratory at the University of Virginia run by Arthur Lichtenberger of the Electrical Engineering Department. HEMT transistors were obtained by NRAO as part of a wafer purchase by JPL. Although room temperature HEMT amplifiers are in relatively common use, the cryogenic-cooled amplifiers designed by NRAO's Marian Pospieszalski outperform them all. His designs are based on a model[8] he developed for cooled HEMTs. They are in use throughout radio astronomy and were key to the success of NASA's Wilkinson Anisotropy Probe. In all but the lowest frequency ALMA receivers, the radiation gathered by the antenna is focused on a SIS junction where it is mixed with a reference frequency to produce a difference called the intermediate frequency. (For the lowest frequency ALMA receivers, the SIS junction is unnecessary.) The intermediate frequency is then amplified by a HEMT amplifier and digitized for transmission to the AOS Technical Building. The SIS mixers are tunerless, that is, they do not require the motor-driven mechanical tuning stubs of previous mixers designs. This feature is crucial to the reliability of the mixers; mechanisms cooled to cryogenic temperatures are prone to failure.

Each ALMA antenna can host up to ten receivers for the detection of radiation in ten bands of the millimeter/submillimeter spectrum. The initial allotment was only eight bands. Bands 1 and 2 are being implemented at the time of writing this book. The bands are matched to the high transparency "windows" in the atmosphere at the ALMA site. Table 9.1 lists the characteristics and fabricator of each receiver band.

The SIS junctions must be cooled to 4 K (−269 C or −592 F) and the HEMT amplifiers to 15 K, to achieve the performance goal of noise generated in the receiver itself to be no more than three times the limit imposed by quantum physics. To accomplish this, the receivers are built into cartridges that are housed in a cryogenic enclosure and cooled by a closed-cycle helium refrigerator. Figure 9.10 shows a pair of Band 6 receiver cartridges and a cryostat with Bands 3–10 installed. Observations can only be made in one band at a time for a given antenna array. The antenna subreflector tilts slightly to direct the signal beam to the selected receiver. The receiver cryostats are installed in the antenna using a modified airplane catering truck and special handling gear.

Local Oscillator System – The local oscillator system[9] is basic to setting the exact frequency of observation. A reference frequency for the receiver mixers is sent from the AOS Technical Building to the antennas via optical fiber as

Table 9.1 *Receiver characteristics*

Band	Frequency (GHz)	Detector type	Partner: fabricator
1	31–45	HEMT	East Asia: ASIAA-led international team: NAOJ, HIAA, NRAO, U. Chile
2	67–90	HEMT	Europe: under consideration
3	84–116	SIS	North America: HIAA (with NRAO SIS mixers)
4	125–163	SIS	East Asia: NAOJ
5	163–211	SIS	Europe: SRON under contract from ESO
6	211–275	SIS	North America: NRAO
7	275–373	SIS	Europe: IRAM under contract from ESO
8	373–500	SIS	East Asia: NAOJ
9	602–720	SIS	Europe: SRON under contract from ESO
10	787–950	SIS	East Asia: NAOJ

Figure 9.10 Left panel: A pair of Band 6 cartridges. The three temperature tiers are separated by two insulating fiberglass cylinders, from 4 K at the top, to 15 K, to 77 K. Right panel: Cartridges for Bands 3–10 installed in the cryo-container. Credit: ALMA/ESO/AUI/NINS, CC BY 4.0.

the difference between two laser frequencies. It is in the range 27–142 GHz. (For the higher frequency receiver bands, this difference must be multiplied by factors of 3, 7, or 9.) The phase stability required for the local oscillator system is very high, and elaborate mechanisms are employed to ensure that the electrical length of the optic fiber does not change. Antenna motion, for example, can stretch the fiber. Changes are monitored and corrected in real time. The early development of this system was part of the design and development program of the MMA, done at NRAO's Tucson site by John Payne and Larry D'Addario.

Intermediate Frequency System – The intermediate frequency system transmits the received signal from the antennas to the correlator in the AOS Technical Building. The signals are digitized at the antennas by high-speed analogue to digital converters (ADCs) before being transmitted on optic fiber cables. An ADC samples the incoming signal 4 billion times per second and delivers a digital output giving the strength of the signal as one of eight levels, encoded in three bits. The ADCs utilize 0.25 µm chip technology developed by STMicroelectronics.[10] Alain Baudry, of the University of Bordeaux, led the team that developed these ADCs, under contract from ESO.

Correlators – The soul of a radio interferometer is its correlator. ALMA has two correlators,[11] a large one for the array of 12 m antennas and another for the ACA, now named the Morita Array.[12] The purpose of the two correlators is the same but their architectures are different. The correlator for the large array is of a conventional architecture whereby signals from all possible pairs of antennas are multiplied ("correlated") to calculate what are called "visibilities." A mathematical operation called a Fourier Transform, performed later in another computer, then produces an image from the visibilities of the object that was

Figure 9.11 View along the rack fronts of one quadrant of the correlator for the 12 m antenna array, located in the ALMA Technical Building at 5,000 m above sea level. Credit: ALMA/ESO/AUI/NINS, CC BY 4.0.

observed by incorporating information about their amplitudes and phases. The correlator for the Morita Array performs these functions in the reverse order, Fourier transforming the data streams before correlation. The Morita Array correlator[13] was designed and built by the Fujitsu Corporation in cooperation with the NAOJ.

The correlators operate at blinding speed. For example, the large array of 50 antennas has 1,225 unique pairs for which the visibilities are calculated by the correlator billions of times per second. The correlator can be considered to be a supercomputer, capable of ~100 trillion arithmetical calculations per second, but it is a very single-minded supercomputer, only capable of this one function. It contains 135,000 integrated circuits on 3,000 printed circuit cards. These consume about 150 kW of power. The heat generated is dissipated by chilled air circulating through the electronics. A supply of spare integrated circuits is available to replace failed units. There are plans to build a new, even more, capable correlator in the coming years.

The correlator[14] for the large array of 12 m antennas was built in the NRAO Central Development Laboratory. It was designed by Ray Escoffier and the project was managed by John Webber as head of the laboratory. The correlator was built in four sections or quadrants, each delivered in succession to the AOS Technical Building and installed by a team led by Rich Lacasse. The first quadrant is shown in Figure 9.11. It was the only major ALMA deliverable to be completed at a cost below the original budgeted amount.

Software

One of ALMA's major software packages is a real-time operating system that controls the array. Large projects often experience cost overruns and delays to completion with their software requirements. This was not the case with ALMA, where the software deliverables only needed a modest budget increase of nine percent to meet their deadlines. Of course, software is never really finished. As the real-time system began to be used, bugs were subsequently revealed and repaired. The rate at which bugs appeared tapered off in time, but improvements to the system continue. The Computing IPT was led by Brian Glendenning. Any of a number of data reduction applications may be used to process ALMA data and produce images. The Common Astronomy Software Applications (CASA) package is one. AIPS, GIPSY, and MIRIAD are other examples. CASA, developed by NRAO, was chosen for the ALMA data reduction pipeline (and for other radio telescope pipelines as well). The pipeline processes

ALMA data automatically, producing so-called reference images that are available via the ALMA Science Archive (ASA). These can be analyzed and published as is, or processed further by the user.

Local Benefits of ALMA

Under the terms of the concession agreement with the government of Chile, ALMA pays for the use of its site. There are three annual payments to make: one to Bienes Nacionales (BN) for rental of the access road, one to CONICYT (now Agencia Nacional de Investigación y Dessarollo, or ANID) for the development of astronomy in Chile, and one to Region II for the benefit of the local communities, principally, the village of San Pedro de Atacama, which includes several other villages in its governance. The average annual amount since the first payments in 2004 is a total of roughly $1 million, divided roughly in half for ANID with the remainder split between BN and Region II. The funds paid to ANID, together with the 10 percent share of observing time on ALMA, along with similar arrangements with the large optical observatories in Chile, have had a profound effect on the Chilean astronomical community. Astronomy departments are now flourishing at universities throughout Chile. ALMA maintains close contact with this community, teaching courses, organizing scientific meetings, and, up to the time of this book, welcoming over 100 interns and postdocs to activities at the JAO headquarters in Santiago. In addition, ALMA has reinforced the education of a larger cadre of technically savvy engineers and astronomers. Although the payments were requirements that emerged from lengthy negotiations for access to the ALMA site, it has been gratifying to see the positive results of the investments. A supportive local community is a great benefit to ALMA.

The ALMA Region II Fund supports social and economic development in San Pedro de Atacama and its sister communities. Projects to be funded are selected by a panel that reviews all submitted proposals. The panel members are the Governor of Region II, the Regional Secretary of Social Development and Family, the Mayor of San Pedro de Atacama, and the President of the Atacameño Community of Toconao. Projects that have been funded range from the small, a tourist map, to the large, equipping a first-aid and dental health center. Science education in the local community is a priority for ALMA. Volunteer time together with funding for teachers and physical improvements to the schools in Toconao have raised the students' scores in national standardized tests. The school in this small, remote village has ranked in the top 100 in Chile. Plans are being made for a science museum in San Pedro de Atacama with a planetarium.

Figure 9.12 The Estancia Barrio ranch has been restored as a museum by ALMA. It lies above the OSF in the elevation band that supports vegetation. Credit: Carlos Padilla; ALMA/ESO/AUI/NINS, CC BY 4.0.

Respect for the local culture has been a hallmark of ALMA's relations with its community. Care has been taken to include religious leaders, officials, and prominent citizens in important events. ALMA has also helped in the preservation of Atacameño's cultural heritage. Working with the Chilean Museum of Pre-Columbian Art, ALMA is investigating the origin and purpose of the "Saywas," archeological mounds of stone that may possibly predict astronomical events through the shadows they project as the sun rises. Figure 9.12 shows the restoration of an "Estancia," a shelter used by local shepherds who grazed animals in the area. It lies in the zone of vegetation above the OSF where the dew is sufficient to support plant growth.

The *Universe of Our Elders* is a 19-page preview[15] of a book to be published on conclusion of a joint ethno-astronomy study conducted by the Universidad Católica del Norte, the Archeological Museum Le Paige in San Pedro de Atacama, and ALMA. The goal of the project is to recover the vision of the Universe held by the inhabitants of the valleys and salt flats below the Andes from San Pedro de Atacama to Ollagüe on the Bolivian border.

Figure 9.13 Chilean President Sebastián Piñera (left) speaking at the ALMA inauguration after being introduced by ALMA Director Thijs de Graauw (right). The screen shows astronomer Antonio Hales on the high site waiting for the command to put the array in motion and begin ALMA operations. Credit: ESO, CC BY 4.0.

Inauguration

ALMA construction was completed on schedule and in time for its inauguration on 13 March 2013. It was a grand affair held at the OSF and attended by more than 500 guests. The guest of honor was Sebastián Piñera, President of Chile, shown while giving his speech in Figure 9.13. In his remarks he said:

> One of our many natural resources is Chile's spectacular night sky. I believe that science has been a vital contributor to the development of Chile in recent years. I am very proud of our international collaborations in astronomy, of which ALMA is the latest, and biggest outcome.

After his address, President Piñera ordered the start of ALMA operations by signaling via direct video link to ALMA scientist Antonio Hales, who was on the high site. Upon getting the green light, Hales contacted the ALMA Control Room via radio and requested that the array be set in motion, pointing the antennas to the center of the Milky Way, all to the theme music from *Cinema Paradiso*. It brought tears to some eyes.

There were many other dignitaries in attendance, including Jorge Molina Cárcamo, the governor of Region II at the time of the negotiations for the ALMA site, and Sandra Berna, the mayor of San Pedro de Atacama. The inauguration posed some vexing protocol issues, but such matters are all in a day's work for ESO which solved them in time for the event. One concerned the flags that fly in front of the OSF Technical Building. It was important that they all be of the same size. Massimo Tarenghi managed to locate and buy them in time. There was also an issue with Taiwan, which Chile does not recognize. The Taiwanese flag was not flown, given the many official Chilean government representatives.

A distinguished attendee of no political office, but held in high esteem by the many astronomers who had come to San Pedro de Atacama over the years, was Tomás Pobleta Alay ("Don Tomás"). His hotel, La Casa de Don Tomás, had hosted astronomical visitors for over twenty years, among them those coming for ALMA and APEX, the Submillimeter Receiver Laboratory group at the Harvard CfA, the Princeton University and Caltech groups studying the cosmic background radiation, and the Cornell University group looking to build a submillimeter telescope on Cerro Chajnantor. Don Tomás was more than welcoming, often entertaining guests at his home. At the

Figure 9.14 Left panel: Tomás Pobleta Alay, owner of La Casa de Don Tomás and friend of astronomy. Right panel: A three-way handshake at the ALMA Inauguration. Left to right: Dick Kurz, Masato Ishiguro, and Bob Brown. All three were critical to the establishment of the division of effort among the ALMA partners in the Bilateral and Trilateral ALMA Agreements. They also played important roles in their own communities – Europe, Japan, and North America – in furthering ALMA and managing the project. Credits: (Left) Paul Vanden Bout; NRAO/AUI/NSF, CC BY 3.0; and (Right) Courtesy of Tetsuo Hasegawa, reproduced by permission.

inauguration, he was eager to shake hands with the President, with whom he shared political views. Figure 9.14 shows a portrait of Don Tomás in the left panel.

And there were those at the inauguration who had conceived, promoted, sought funding for, and built ALMA over decades. For them, it was the realization of a dream and a highly emotional event. Figure 9.14, in the right panel, shows three of these individuals, the project managers/directors from the three ALMA Executives in a three-way handshake.

Notes

1 A summary technical description of ALMA can be found in Wootten and Thompson (2009). See also Iguchi et al. (2009) for a description of the ALMA Compact Array (ACA).
2 The full text of Giacconi's speech can be found at: https://library.nrao.edu/public/memos/alma/misc/ALMAU_3.pdf.
3 The contract from Santiago to FIU was made in agreement with AURA, taking advantage of their established path pertaining to optical telescope initiatives – present, and future – in Chile. Data transport via FIU is coordinated by NSF initiatives known as AmPath and AmLight.
4 To see an antenna transporter in action, watch the videos The ALMA Transporter Garage, www.youtube.com/watch?v=DBu9k1eq4HU, and Hauling an ALMA Telescope, www.youtube.com/watch?v=Ss0bxoLsOUs.
5 The section of the ALMA Operations Plan describing the *Sistema de Turno* in more detail can be found at: https://safe.nrao.edu/wiki/pub/ALMA/ARCinCSVrules/ARCinCSV.pdf.
6 Invar is an alloy of 36 percent nickel and 64 percent iron that is known for its low linear coefficient of thermal expansion. It has about 10 times less linear thermal expansion per degree Celsius than steel. Invar stands for invariable, in reference to this quality. The Invar cone in the Vertex antennas helps shield the reflector BUS from thermal distortions in the receiver cabin.
7 The Jansky is defined as 10^{-26} W m^{-2} Hz^{-1}.
8 The European Microwave Association awarded Pospieszalski its Pioneer Award in recognition of the impact his HEMT model has had on the field of microwave engineering.
9 ALMA's photonic local oscillator system is described by Shillue et al. (2012).
10 A technical description of the ALMA ADCs has been given by Recoquillon et al., in ALMA Memo #532 https://library.nrao.edu/public/memos/alma/memo532.pdf.
11 The performance highlights of the two ALMA correlators are discussed by Baudry et al. (2012).
12 The Morita Array is named for Koh-Ichiro Morita, a distinguished Japanese scientist and JAO staff member who tragically died from a head injury received during a mugging near his home in Santiago on 12 May 2012.
13 A technical description of the Morita Array correlator was given by Abe, Tsutsumi, and Hiyama (2014).

14 A technical description of the ALMA large array correlator was given by Escoffier et al. (2007).

15 The preview of *The Universe of Our Elders* can be found at: https://almaobservatory.org/wp-content/uploads/2016/11/alma-etno_2013.pdf

See also a brochure (in Spanish) describing the ALMA Fund at: www.almaobservatory.org/wp-content/uploads/2022/07/Informativo-Fondos-ALMA-2022.pdf.

Promises Fulfilled

"The scientist does not study nature because it is useful;
she studies it because she delights in it,
and she delights in it because it is beautiful."

Henri Poincaré (paraphrased)

ALMA Science Goals

After five years of telescope operation, the ALMA Board formulated a working group to assess the progress toward answering the scientific goals that had established the need for ALMA in the first place. This working group prepared the ALMA Development Roadmap, a vision for the future of the telescope, building on results from its first five to ten years of operations. The Roadmap[1] reported in 2018 that three of the principal science goals had been accomplished, namely:

- *The ability to detect spectral line emission from CO or C^+ in a normal galaxy like the Milky Way at a redshift of $z = 3$, in less than 24 hours of observation;*
- *The ability to image the gas kinematics in a solar-mass protoplanetary disk at a distance of 150 pc, enabling one to study the physical, chemical, and magnetic field structure of the disk and to detect the tidal gaps created by planets undergoing formation;*
- *The ability to provide precise images at an angular resolution of 0.1 arcsecond.*

In contrast to the ALMA Roadmap, the CAA report commissioned to assess the impact of a reduction in the number of antennas on ALMA science had concluded that it would not be possible for ALMA to achieve any of the three science goals with only 50 antennas. This chapter will take a closer look at these proposed science goals using specific examples from the roughly 4,000 projects in the ASA at the time of the writing of this book.

Molecular Gas in Distant Galaxies – Further discussion of the first goal was made in an Astro2020 white paper[2] on "Activities, Projects, and State of the Profession Considerations." According to the white paper authored by Crystal Brogan, the first science goal "*has been achieved in spirit, though not in detail – indeed, a major achievement of modern astrophysics is the realization that 'Milky Way-like' galaxies do not exist at z = 3.*" That is, the goal was unachievable by definition. The progenitors of a normal galaxy like the Milky Way do exist at $z = 3$ and higher redshifts, and galaxies with the same mass in *stars* as the Milky Way have been detected in CO emission at high redshift. Two early ALMA projects, led by Fabian Walter at the MPIA in Germany and Manuel Aravena at the Universidad Diego Portales in Chile, made these detections, which led to more than 10 publications and were the building blocks for the ALMA Large Program known as ASPECS (ALMA SPECtroscopic Survey in the Hubble Ultra-Deep Field).[3]

The question of whether ALMA met this goal also depends on the definition of "Milky Way-like." If defined as a galaxy at $z \sim 3$ with the same mass in interstellar gas as the Milky Way, then this has not yet been demonstrated due to the time required for such observations, although it would be less than 24 hours. An even more impressive demonstration of ALMA's power would be to make an image of the CO distribution in a high redshift galaxy that resembles the Milky Way. This would require at least 100 hours of telescope time,[4] a proposition that some astronomers consider "high risk, high reward," but is yet to be done.

Protoplanetary Disks as Sites of Forming Planets – The source HL Tauri emerged as a poster child for the goal of studying the environment of a nearby star in formation and imaging its protoplanetary disk. HL Tauri is a million-year-old Sun-like star located approximately 450 light years from Earth in the constellation of Taurus. It was observed as a test target in 2014, to assess the functionality of the ALMA long baseline capabilities, and several times since then. The first HL Tauri image (Kwon et al. 2011) was obtained using CARMA with baselines up to 1.5 km. The image, shown in Figure 10.1 (bottom panel), revealed a structure in the disk that hinted at early signs of planet formation. The structure had been invisible to previous images, taken with telescopes lacking the resolution of ALMA. See Figure 10.1 (top panel) for an example. The detections were of the emission from the dust, rather than the gas, in the disk surrounding this source. The paper[5] that presented the iconic image of HL Tauri has garnered more than 1,000 citations and counting, making it one of the most prolific ALMA results to date. More than two dozen papers utilizing this verification dataset, and complementary data, have been published since. *"When we first saw this image, we were astounded at the spectacular level of detail. HL Tauri is no more than a million years old, yet already its disc appears to be full of forming planets. This one image alone will revolutionize theories of planet*

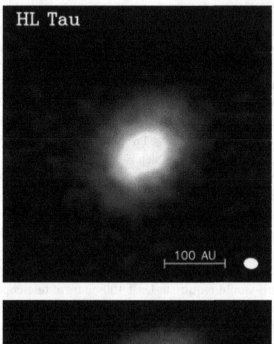

Figure 10.1 Top panel: Image of the protoplanetary disk surrounding the young star HL Tauri, with contours indicating the signal strength, as detected by CARMA before ALMA was built. Bottom panel: The ALMA image reveals the structure of the disk. The gaps are possible locations for planet formation. Credits: (Top) Courtesy of Woojin Kwon, reproduced by permission; (Bottom) ALMA Partnership, et al. (2015); ©AAS, reproduced by permission.

formation," explained Catherine Vlahakis, who worked at the JAO in Santiago at the time of the observations.

Observations of HL Tauri and other similar sources set the stage for ALMA Large Programs such as the Disk Substructures at High Angular Resolution Project (DSHARP), and later the Molecules with ALMA at Planet-forming Scales (MAPS). The DSHARP program focused on studying the dust and gas (CO) in a sample of twenty relatively nearby, bright, and large protoplanetary disks. The stunning results were presented in a special focus issue[6] of *The Astrophysical Journal Letters*, and showed the distinct and complex shapes and structures at the earliest sites of planet formation. The authors interpreted the observations as evidence for unseen planets interacting with the dust

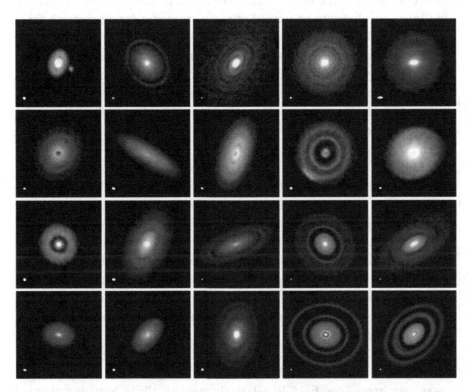

Figure 10.2 Twenty protoplanetary disks imaged by the DSHARP program. This is a groundbreaking gallery of diverse shapes and structures apparent in early stages of star formation, likely pointing to planets forming earlier in the process than previously thought. The small white ellipses in the lower left corner of each panel represent the resolution – or level of discernible detail – of the observations, showing that very minute structures, just about five times the distance from the Earth to the sun, are able to be studied in these images. Credit: Andrews, et al. (2018); ©AAS, reproduced by permission.

and gas around young stars. This survey confirmed that HL Tauri was not a unique case, but rather, that there were signs of planet formation in nearly all disks observed in great enough detail, based on the concentric gaps and narrow rings that were seen as substructures in the light that was detected. The collection of diverse, spectacular protoplanetary structures, shown in Figure 10.2, was only beginning. It is obvious that ALMA has met its second science goal.

ALMA's Impact on My Career

ALMA has changed the lives of many astronomers in the world, including in Chile. I was lucky to be trained in radioastronomy during research for my master's degree in Chile and for my PhD in France. That gave me the tools to be ready when ALMA was inaugurated in 2013, the same year I graduated. I was fortunate enough to be awarded ALMA observing time during my first postdoc in the United States, which helped kick off my career. I became a frequent ALMA user. I moved back to Chile at the end of 2016 to work at the ALMA headquarters in Santiago, where I was able to learn even more about the telescope.

ALMA has revolutionized the field of planet formation, but also the field of astrochemistry. Actually, ALMA has become the astrochemist's best tool, allowing us to detect faint lines in regions we could not see before, such as protostars and protoplanetary disks. For example, we discovered the first complex molecule and made the first measurement of the ratio of nitrogen isotopes in a disk. After the famous HL Tau image appeared, I had the privilege to participate in the DSHARP project which revealed incredible dust substructures in disks. More recently, I was co-leader of the MAPS project, where we mapped the emission from tens of different molecules in disks.

ALMA will continue to make exciting discoveries, and I feel lucky to be in a position where I can make a contribution to the field of astrochemistry. This field is relatively new in Chile but now there are a few of us across the country. More importantly, since I became a faculty member at the Pontificia Universidad Católica, I have been given the chance to start training the younger generation in the field of astrochemistry and planet formation. My hope is that Chile will become a leader in these fields, and take advantage of the fantastic observatories we have in our country.

For all these reasons, ALMA means so much to me, not only because it gave me my career, but also on a personal level. I actually grew up very close to the ALMA site, in the city of Antofagasta. I moved to Santiago in 2004 to enroll in the Universidad de Chile, the same year that ALMA construction started. However, back then I never imagined I would become an astronomer and use this fantastic telescope that is now practically in what was once my backyard.

Viviana Guzmán
Pontificia Universidad Católica de Chile
Santiago, Chile

High Angular Resolution Imaging – ALMA's performance exceeded the angular resolution defined in the third science goal by a factor of 10, achieving an angular resolution of 10 milli-arcseconds, or 0.010 arcseconds. This is equivalent to discerning a basketball hoop located about 9,000 km distant (the distance between Los Angeles and Munich, Germany), or even a common whale at the distance of the Moon. It is clear that this third science goal was critically important to obtaining the images of protoplanetary disks, but it applied equally to a broad range of astronomical studies. It was a breakthrough in high angular resolution imaging. Now, after nearly a decade of observing with ALMA, imaging with sub-arcsecond angular resolution has become rather commonplace. More than 650 projects in 4,000 have specified at least one aspect of their observing campaign to observe with greater detail than 0.1 arcseconds. That is, about one in six ALMA projects are aiming at this level of detail when imaging the sky.

The physical size of an object that appears to cover 0.1 arcsecond on the sky depends on the distance to that object from the telescope on Earth. At the distance of some of the nearest sites of star and planet formation, that distance equals about 10 times the distance of the Earth to the Sun, or 10 AU. An observation with better than 0.1 arcseconds resolution toward our neighbor galaxy, the Large Magellanic Cloud, would resolve details of about 5,000 AU. That is significantly larger than the size of our solar system, but much smaller than the size of the clouds where groups of stars are formed. Hence, one can begin to study the sites of planet formation within our Galactic neighborhood, and star formation beyond the limits of our own Galaxy. Even phenomena in the seemingly distant universe, such as massive, energetic jets launched from active galactic nuclei of galaxies born millions of years ago, can be studied in detail.

In order to have an idea of what it takes to obtain this angular resolution goal, it is helpful to revisit the different array configurations available at ALMA. The ALMA antennas can be rearranged in a variety of positions, and the farthest distance between any two antennas determines the smallest level of detail that can be observed. Depending on the exact wavelength being observed, 0.1 arcsecond resolution can be obtained with the configuration whose longest baseline is almost 800 m, or half a mile. If observing with a longer wavelength, then the antennas would have to be spread even farther, possibly up to about 8 km in order to detect such small structures. Nonetheless, ALMA was designed to operate with its antennas separated by up to 16 km, which is to say that the goal of observing with 0.1 arcsecond resolution is well within its design reach, as long as the additional "precision" requirements are also met.

Long baselines are not enough to meet the "precision" aspect of the third science goal. It is worth delving into the meaning of "precise." For astronomical imaging with ALMA, this was (precisely) specified: "*Here the term precise image means representing within the noise level the sky brightness at all points where the brightness is greater than 0.1 percent of the peak image brightness. This requirement applies to all sources visible to ALMA that transit at an elevation greater than 20 degrees.*" In other words, synthesizing the three science goals, it was imperative not only to see objects that were relatively small and very distant, but also to distinguish a vast array of subtle differences in source structures and the different combinations of gas and dust that reside therein. The study of minute features in distant objects depends on corrections of signal fluctuations due to instrumental and atmospheric effects in order to reach what's known as a high "dynamic range" image. With very careful signal processing, the goal is to detect the two extremes of faint and bright light in the same observation – for example, a booming central emitting region of an object, as well as the faint outer limits. The complete picture is critical to a deeper understanding of complex astronomical targets.

The result of such stringent constraints on antenna configurations and imaging capacity is stunning pictures of astronomical objects in unprecedented detail. Astronomers began showing detailed color-scale ALMA images at conferences and in their publications, rather than the contour plots, spectra, or other less eye-catching data representations they had used when observing with previous arrays. In the early years of ALMA observations, conference presentations regularly included a "then" and "now" view of a certain astronomical object under study, emphasizing how much more detail could be seen in the new view with ALMA. Speakers would use phrases like, "*And that observation was made in less than an hour of observations with ALMA,*" or "*That was done in less than half the time of previous observations,*" emphasizing the impressive power of ALMA to observe great detail within a reasonable amount of time.

The Breadth of ALMA Science

In the approximately 10 years since ALMA operations began, scientific teams worldwide have dreamt, planned, executed, and analyzed observations addressing the variety of scientific topics incorporated into the ALMA design concept. From cosmology to neighboring galaxies to our own Milky Way galaxy and the stars, planets, and other astronomical bodies within, here in Figure 10.3 we provide a snapshot of some of the impactful discoveries in ALMA's first 10 years of operations.

ALMA's First Image – With just 12 antennas working together on the Chajnantor Plateau, ALMA observed the Antennae, a pair of colliding galaxies, and the result became the first scientific image from ALMA to be released on 3 October 2011. This is just one example of a new view of the Universe, beyond what had previously been studied with visible and infrared telescopes at similar resolution, or with radio telescopes at lower resolution. The galaxies in the Antennae began colliding relatively recently, only a few hundred million years ago, making it one of the youngest known major galaxy mergers. As such, at a distance of about 70 million light years, telescopes can discern the two remaining nuclei from the original galaxies, dusty filamentary lanes that connect these

Figure 10.3 A word cloud representing the abstracts of publications using ALMA data as of September 2022. Very roughly, the size of the word corresponds to its frequency in the text of the publication abstracts, summaries of the work. The image gives a visual impression of the topics that dominate ALMA science programs. Credit: Courtesy of Felix Stoehr, reproduced by permission.

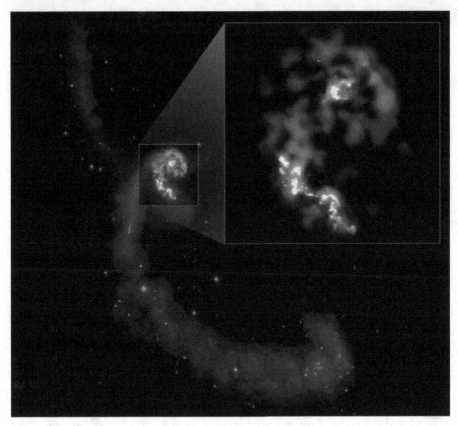

Figure 10.4 The first scientific image with ALMA, of the two interacting galaxies that form the Antennae. The image is a composite of CO emission from ALMA, H I emission from the VLA, and optical/infrared emission from HST and CTIO. Credit: ALMA/ESO/AUI/NINS, CC BY 4.0; HST/NASA/ESA, CC BY 3.0; J. Hibbard; NRAO/AUI/NSF, CC BY 3.0; NOAO/AURA/NSF, CC BY 4.0.

nuclei, and star clusters scattered throughout. ALMA observed the dust and the carbon monoxide gas that emit millimeter-wavelength light in the Antennae galaxies, and soon thereafter the Observatory produced its first image, shown in Figure 10.4, to celebrate ALMA's first science result. The ALMA data were obtained as part of the Science Verification campaign, and they were compared with similar observations from other millimeter wavelength arrays to corroborate ALMA's first observing modes.

It was a Chilean graduate student, Cinthya Herrera, who published the first peer-reviewed journal article with ALMA observations while completing her doctorate degree in Paris, France.[7] In the paper, she and her co-authors combined ALMA data of the Antennae with observations they had made with the VLT, an optical facility also in the Atacama Desert, about 300 miles away from ALMA. This allowed them to compare the carbon monoxide and molecular hydrogen gas, both important components in the structures that form stars.

It is worth reiterating here that carbon monoxide is one of the main design drivers and likely the most commonly observed molecule with ALMA, as it allows astronomers to peer into the otherwise invisible hydrogen clouds, which are the sites where stars form. In fact, carbon monoxide is targeted in the observing setup of about one-third of projects that were observed with ALMA to date. This has been the "workhorse" molecule for ALMA since inception. Whereas the study of CO began with single-dish telescopes, ALMA was honing in on greater detail in the morphology and kinematic structure of CO gas reservoirs throughout the Universe. The first detection of CO gas around a young exoplanet orbiting a star about 400 light years away came in 2022, when Jaehan Bae[8] found this small but important detail in the data that were part of the ALMA Large Program known as MAPS. Astronomers had expanded upon the goal to detect gas structures around stars and were endeavoring to study the gas that goes on to specifically form planets in those systems.

Astrochemistry – Going beyond CO, astronomers have used ALMA to study the universe as a chemistry lab, searching for molecules much more complex than carbon monoxide. Primarily, the search is on for increasingly complex carbon-based molecules, as these are considered the basic building blocks of life. If the conditions are right, simple molecules like CO, hydrogen, and water can go on to form amino acids and proteins, which are notably important to human biology and life in the Universe. Astronomers search for environments reminiscent of what are likely the precursors to planetary systems that could host Earth-like planets. One extremely young star with characteristics to suggest that it will go on to become a star similar to our Sun is known as IRAS 16293-2422, located at a distance of about 400 light years in the constellation Ophiuchus. It has been studied extensively with ALMA[9] since first being observed in 2012, and it was found to include the simple sugar glycolaldehyde, as well as methyl isocyanate, which is lethal but can form peptides and amino acids by combining with other molecules in interstellar environments. Figure 10.5 shows the spectrum of one of eight star-forming clouds in the galaxy NGC 253, about 25,000 times farther away than IRAS 16293-2422. A total of nineteen molecular species were identified in this cloud. Detections of complex molecules like these inspire theories of how life as we know it may have begun throughout the nearby Universe.

ALMA does not only look at stars similar to the Sun, but also the Sun itself. Teams have proposed solar observations since 2016, and data were released to the astronomical community in 2017 as part of the Scientific Verification

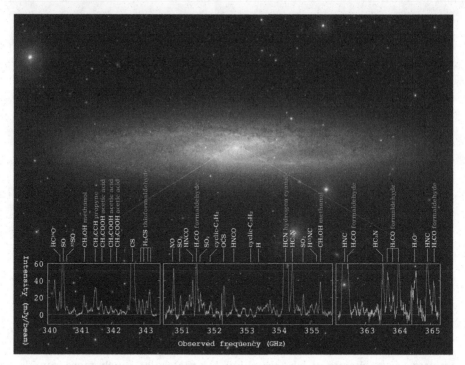

Figure 10.5 The spectrum of molecular emission lines detected in the heart of the star-burst galaxy NGC 253. Credit: J. Emerson; ESO, CC BY 4.0; ALMA/ESO/AUI/NINS, CC BY 4.0; adapted from Ando et al. (2017); ©AAS, reproduced by permission.

process[10] to demonstrate ALMA's capabilities in this area. ALMA is able to probe the solar chromosphere, and it is important, especially for determining the temperature of the chromosphere in high detail. The chromosphere is one layer of the Sun's atmosphere, lying between the photosphere below and the corona above; the chromosphere is likely important in transferring heat from the solar interior to the outermost layer, the corona. The ALMA observations are often analyzed in concert with data from other telescopes that record other wavelengths of light, including NASA's Solar Dynamics Observatory and the Interface Region Imaging Spectrograph. Interest among an international cohort of solar researchers can be summarized in the statement by Maria Loukitcheva who has been studying millimeter waves from the Sun for over two decades[11]: *"The importance of ALMA for solar physics is indisputable."* Sven Wedemeyer, a professor at the University of Oslo who leads several major research initiatives to observe the Sun with ALMA, reinforced the importance but also the challenge, stating: *"Interferometric observations of a dynamic source like the Sun and the reliable reconstruction of corresponding image series are challenging tasks."* The observed imprints of magnetic loops and variations in temperature over time are important phenomena to better understand the solar composition.

Distant Galaxies – ALMA has been on the hunt to detect the most distant galaxy, as it would appear when it was very young, formed early on in the history of the universe. Astronomers use observations of distant galaxies as "time machines" in order to understand the conditions of the universe at an earlier time. One galaxy, named SPT0418-47, was observed by ALMA and established as the most distant then known Milky Way look-alike. What was remarkable about this galaxy, about 12 billion light years away, was how unremarkable it looked even though it was formed when the Universe was just 1.4 billion years old. The results published by Francesca Rizzo,[12] at the time a graduate student at the Max Planck Institut für Astrophysik in Germany, showed that it has two features in common with our Milky Way – a bulge and a rotating disk. The galaxy appears surprisingly unchaotic, even though it was previously expected that galaxies in the early universe should be turbulent and unstable, eventually becoming more orderly with time. The observations of SPT0418-47 utilized a clever technique and an effect known as gravitational lensing, effectively converting a nearby galaxy into a powerful magnifying glass that allows ALMA to see unprecedented details of objects in the more distant universe. In this case, the intense gravitational pull of a galaxy between ALMA on Earth and SPT0418-47 focused the light of the distant galaxy, and as a result, SPT0418-47 appeared magnified as a near perfect ring of light around the nearby lensing galaxy. A model of the lensing allowed the team of astronomers to reconstruct the galaxy's true shape, which turned out to have a galactic disk that was much more orderly than any other known distant galaxy.

Less than one year later, astronomers published a study of another galaxy BRI 1335-0417,[13] whose mass is roughly equal to the Milky Way, and whose distance is similar to that of SPT0418-47. Takafumi Tsukui, a graduate student at the Graduate University for Advanced Studies (SOKENDAI) in Japan, led the study of an image they found in the ASA that revealed evidence of a spiral structure, in addition to the rotating disk and massive central region. The spiral structure was therefore the most distant such example known at the time, and Tsukui expressed his excitement saying *"The quality of the ALMA data was so good that I was able to see so much detail that I thought it was a nearby galaxy."*

ALMA is not limited to studying galaxies like the Milky Way with notable spiral structures. Astronomers continuously search for faint signals, thanks to ALMA's high sensitivity, that are likely candidates for even more distant galaxies. To this end, ALMA observers found two galaxies that formed more than 13 billion years ago, less than 1 billion years after the Big Bang. The international team of astronomers had targeted forty very specific galaxies at "cosmic dawn" as part of an ongoing large program called Reionization-Era Bright Emission Line Survey (REBELS). The results[14] suggest that there may be more galaxies forming in early in the history of the Universe than expected.

Competition for Observing Time

ALMA's scientific payoff is clear from the published results, but it is worth elaborating on the observing process. ALMA is truly a community-driven, publicly funded, and internationally run facility. How does one make ALMA observations, in practice? Next, we explain the process from idea to observation to discovery to publication of scientific results.

Open Skies – ALMA was designed for a breadth of scientific topics, beyond the specific examples we just discussed. ALMA scientific categories include, among others, cosmology and the high redshift Universe, galaxies and galactic nuclei, interstellar medium, star formation, astrochemistry, circumstellar disks, exoplanets, the solar system, stellar evolution, and the Sun. Anyone in the world can propose an observation with ALMA, at no cost to the proposer. According to the Trilateral ALMA Agreement, the time allocated for observation with ALMA should, over the long term, adhere to the relative level of funding of the respective regions, resulting in 33.75 percent for the North American partners, 33.75 percent for ESO member states, and 22.5 percent for the East Asia. Taiwan is a unique case whose observing time can be tallied as North American or East Asian, according to their agreements with both regional members. Additionally, 10 percent of time is allocated for Chilean proposals, in exchange for their role as host of the telescope. Finally, proposals can be submitted by astronomers outside of the previously mentioned regions and alliances. Such "open skies" time is divided among the partners according to their respective shares for up to five percent of ALMA's observing time, and beyond that threshold, it is charged to the North American share.[15]

In addition to the annual call for proposals, up to five percent of the observing time in a Cycle can be allocated by the ALMA Director as Director's Discretionary Time. These proposals are generally cutting-edge science, typically those for which immediate observations are needed to capture an unanticipated short-lived astronomical event or science cases that have the potential to lead to a breakthrough discovery. In some cases, the proposed observations may be rather risky, but could have a large impact from only a small amount of observing time.

Scientific assessments by reviewers (detailed in following sections) form the foundation for what is known as a "scheduling exercise" to determine which top ranked proposals could realistically be executed in the upcoming year. Final grades of A (top priority), B, C, and U (unscheduled) are assigned to each proposal, and proposal authors are also notified of the "quartile" within which the proposal ranked. The scientific ranking is not the only determinant of whether a proposal is eventually scheduled, because other factors such as regional balance and scheduling feasibility have to be taken into account.

Annual Observing Cycles – As construction progressed, astronomers worldwide knew that ALMA was coming, and would open up a window to astronomical details they had never seen before, thanks to the combination of ALMA's angular resolution and sensitivity. The first call for proposals was opened in 2011, for what was known as "Cycle 0." Cycle 0 would span 9 months and offer 500–700 hours of time on a somewhat limited array. This period was known as "Early Science," and the observing time came with the risks that accompany early operation of a telescope. Nonetheless, astronomers eagerly applied to use ALMA for their proposed scientific endeavors, since ALMA, though incomplete, was already more powerful than other interferometers that observed at the same wavelengths. Although its formal inauguration would be in 2013, the revolution with ALMA had begun. The "Early Science" period included Cycles 0–2, with each cycle lasting between 9 and 17 months. With the call for Cycle 3 in 2015, ALMA entered into a phase dominated by scientific observations, but still balanced with engineering, maintenance, and development activities. From then on, the cycles lasted twelve months each, following a cadence such that the call for proposals closed in April of each year, for observations to be scheduled between 1 October of the same year, and 30 September of the following year.

With each cycle, more capabilities were offered, notably more antennas, more receiver bands, and more available observing time. The available observing time eventually reached 4,300 hours (about half the hours in a year) on the 12 m array in Cycle 7, which was a requirement for full operations, and remained at that level as of the writing of this book. Additionally, Cycle 7 offered a "Supplemental Call" due October 2019, in order to add additional projects to the queue, specifically those that could be observed with the ACA during undersubscribed allocations of the observing queue. The main call that year offered 3,000 hours of time each for the ACA, also known as the Morita Array, and Total Power Array. A supplemental call would also be offered in 2021. A smaller one had been offered in Cycle 4, but none are planned for future cycles. At the time of the writing of this book, the most proposals submitted for any single cycle were 1,836 in Cycle 6, steadying between 1,700 and 1,800 for Cycles 7–9. It is worth noting that considering the time requested, ALMA maintains an over-subscription rate of around six, varying somewhat by region and cycle. In other words, for every proposal accepted to be scheduled in the coming year, about five are rejected or unlikely to be observed.

Proposal Preparation – ALMA was designed such that non-experts in millimeter wave interferometry would have the opportunity to pose scientific questions and undertake the corresponding observations. Around the time of the annual call for proposals, community workshops are organized by the ALMA Regional Centers (ARCs), with ALMA staff available to assist astronomers from their region with proposal preparation. In the case of the North American ARC, the ALMA Ambassadors program trains postdocs and graduate students from institutions in North America and Chile, in order for them to subsequently host workshops for their local astronomy communities.

In order to apply for ALMA time, astronomers must prepare a four-page document detailing the scientific justification for the observations; two additional pages are allocated for the largest subset of programs, which also require a team management plan. Additionally, some predefined details of the observational setup must be entered into a software interface known as the Observing Tool. There, the astronomers indicate the coordinates for their observation(s), frequencies to tune the receivers, and required level of detail needed for a detection. The Observing Tool will calculate the amount of time needed to accomplish the goals based on the technical setup. In early cycles, with limited time available, observations shorter than a few hours were encouraged, but with subsequent cycles that time recommendation has been lifted. Astronomers now request the amount of time necessary to accomplish their scientific goals, measured in terms of observing sensitivity to ensure detection of their designated source(s). Moreover, by 2020 advisory committees were encouraging proposals with larger time requests, broader in scope. By Cycle 9, the median request for observing time with the 12 m array was 12.2 hours per project.

Panel Review – Reviews of proposals by panels of experts are common in astronomy and other scientific fields as a mechanism for selecting the most compelling scientific pursuits in a given area for a given period of time. ALMA followed suit for at least Cycles 0–9, for all or a subset of proposals. The panelists were selected for being experts in the field of millimeter interferometry and the scientific topics being studied by ALMA. In addition to expertise, the panelists reflected the regional balance of ALMA. Approximately two-thirds of the panelists were, in equal parts, North American and European, complemented by colleagues from East Asian institutions, and about 10 percent from Chilean institutions. To combat potential biases, intentional efforts ensured gender balance among the panelists as well.

Panels were organized according to scientific category, with the number of panels proportional to the number of proposals received in a given category. The number of panels had to gradually increase to keep up with the growing number of submitted proposals per cycle until leveling off at around 1,800 proposals, in an effort to keep the number of proposals reviewed by a given panelist within a reasonable range. At its peak, panelists of certain expertise categories sometimes read more than 100 proposals, and up to 158 panelists were recruited for a given cycle. Panelists generally committed to serving three cycles before rotating off. It became a badge of honor to be invited, and the panelists often celebrated at a culminating banquet in which they were honored for their service.

Distributed Peer Review – Although panel reviews had been the standard for telescope time allocation at ALMA and other observatories, ALMA instituted an alternative known as distributed peer review, in the hope that it would distribute the burden of the process more equitably. In a distributed peer review process, someone who submits a proposal agrees to be responsible for reviewing a fixed number of proposals among the collection of submitted proposals. The same individuals submitting proposals are also contributing to selecting the best proposals to be observed.

The first implementation of the distributed variant at ALMA was for the Supplemental Call of Cycle 7, in which observations using the ACA could be proposed before 1 October 2019. Several hundred proposals were received in this call, with an equal number of "designated reviewers"[16] participating in the first distributed peer review of its size for astronomy. Thousands of reviews were therefore processed, 10 for each submitted proposal.

Following success in the Cycle 7 Supplemental Call, ALMA decided to ratchet up the magnitude of the distributed process. In the call for proposals of Cycle 8, held in 2021, only the proposals requesting more than 25 hours on the 12 m array or 150 hours on the 7 m array – so-called medium to large proposals – were evaluated via the usual peer review process. In contrast, 1,497 proposals were processed through the distributed panel review process, followed by 1,729 proposals in Cycle 9, the most for any telescope review of its kind to date.[17] These were the first instances in which the majority of observing time at a major astronomical facility had been allocated this way, and an example of ALMA operating at the forefront of the field.

Dual Anonymous Review – Another important aspect of the ALMA observing procedure has been the aim to equitably allocate observations among the worldwide partnership. Following the described proposal review process, certain systematics became apparent in early cycles. There were indications that proposals with a male principal investigator fared better than those led by a female, North American and European proposals were scored higher than those written by authors in East Asia or Chile, and astronomers' success seemed to be correlated with the number of times that they had previously submitted proposals for observations with ALMA, a so-called "prestige bias."

The proposal review process evolved with time, first making the panel reviewers aware of unconscious biases and hiding some author information in the proposal tools, then randomizing the order of authors' names in the cover page, and finally taking the leap to a fully dual anonymous procedure. Ultimately, the proposals had to be written such that the identities of the proposing teams were unknown to the reviewers, just as the reviewers' identities were unknown to the proposing teams at the time of the proposal call. The first fully anonymous ALMA proposal cycle took place in 2021, and by that time many of the major international telescope facilities had adopted this modality as well.

A thorough report investigated the impact that the dual anonymous process had on apparent systematics.[18] The prestige bias seems to have subsided, and PIs who submitted a proposal for the second time had ranks comparable to those most experienced PIs with many years of proposals under their belts. Interestingly, no significant differences of ranks based on gender were found, and the systematics of regional affiliation remained similar to prior cycles. Throughout, the ALMA project has placed high importance on equity and fairness in granting observing time, and the dual anonymous and distributed peer review components of the proposal process were steps in this direction that ALMA continues to monitor for optimal outcomes.

Steps in Remote Observing and Reduction of Data

Remote Observing – The actual process of observing with ALMA, as experienced by the astronomers who proposed the approved observations, is rather mundane in comparison with the heroic adventures of the groundbreaking astronomers who had scouted the ALMA site and embarked on the many negotiations to make ALMA a reality. Once a proposal is accepted for observation, the proposal is translated into an observing setup via the Observing Tool, with the assistance of staff members at the ARCs and the JAO. Staff known as contact scientists are assigned to each proposal, and act as a sort of customer service representative throughout the process. The proposal authors check to make sure all the details are as intended. Then the observations are put into a queue.

Queue observing is a mode by which the projects that meet certain criteria (weather, antenna configuration, orientation of the sky) during a given period are accordingly identified and able to be fit like puzzle pieces into the telescope agenda. Ultimately, observations are done by ALMA staff, rather than the astronomer(s) who proposed the observations. This is certainly an efficient, if perhaps anticlimactic way for the proposing team to obtain data. A team of one or more ALMA astronomers act as Astronomer(s) on Duty together with one or more telescope operators, usually engineers by training, at all times. The observations may be undertaken 24 hours per day, 365 days per year, although the facility is generally closed in February for maintenance and weather constraints, and time is shared throughout the weeks between science operations, engineering, and maintenance of the array. In order to maintain staffing levels, astronomers and operators serve on rotating 8–10-hour shifts throughout the day (and night), with a shift usually lasting a total of 8 days at a time.

The astronomers running the observations at the ALMA site are about 30 km removed from the plateau where the signal meets the ALMA antennas. This is because the astronomers, and other staff, work from the Operations Support Facility (OSF) during their shifts, located at the elevation of 2,900 m above sea level, compared to the more than 5,000 m elevation of the AOS facility at the Chajnantor Plateau where the antennas are situated.[19] The astronomers and telescope operators can control the antennas from the control room in the technical building at the OSF. A photograph of nightshift astronomers in the ALMA control room during early science observations is shown in Figure 10.6. The observed astronomical signal passes through a fiber optic cable buried underground from the AOS to the OSF, on its way to Santiago and around the world.

Since late 2019, ALMA engineers and astronomers have developed the tools to enable operating the antenna array from an even greater distance, via the control room extension at the ALMA Santiago Central Office (SCO). In other words, astronomers can run observations from over 1,600 km away from the antennas themselves. While indeed this was a very convenient and critical feature during certain periods of the COVID-19 pandemic, the control room extension had already been in planning phases since mid-2019, with the expectation that it will endure even as COVID-related travel restrictions and safety protocols were lifted. The control room extension ensures accessibility, flexibility, and reliability for the observing process while reducing the travel budget implied by the previous mode of operations. While some staff will inevitably always need to travel to the antenna site, it is no longer necessary for all staff to travel for every shift.

Data Reduction – Once the data come off the telescope, and follow their fiber optic path to Santiago, they are still not ready for scientific investigation. The JAO, together with the ARCs, has taken responsibility for the next stage of data quality assurance and reduction.[20] The quality assurance ensures that the

Figure 10.6 Nightshift in the ALMA Control Room during early science. Left to right: (standing) Rainer Mauersberger (ESO), Robert Lucas (ESO), Alison Peck (US, Deputy Project Scientist), Mareki Honma (Japan); (seated) Adele Plunkett (US); Manuel Aravena (Chile). Credit: Max Alexander; ESO, CC BY 4.0.

data meet the specifications set forth by the project's Principal Investigator. The data processing and reduction is performed via software known as the ALMA Science Pipeline, often but not always without human intervention. In the process, signals are calibrated and reference images of the target and calibration sources are generated. The principal investigator will receive the raw data and some or all requested images, along with access to the calibrated data, in order to begin their scientific analysis, or to perform more specific and nuanced data processing.

ALMA Science Archive – Data are available to the astronomical community via the ASA, accessible via a webpage. Behind the web interface is a database that stores all of the data, together with information about the observations that generated these data. The Principal Investigator (PI) will initially access their requested data via the ASA, following a specified link or by logging into the system and navigating the search and download functions. ALMA adopted a proprietary period of one year for most projects, meaning that datasets are exclusively reserved to the proposing team for the first twelve months after the data are released to the PI, that is, after quality assessment and pipeline reduction. The PI may delegate other users to access the data. Anyone can learn of the existence of specific data from the time that they are made available to the PI, with the data having been ingested into the archive database, and the date on which the data will be made public is readily apparent. Beginning at one year after the time of data release, the data are completely public, and accessible to anyone, consistent with NSF's Open Skies policy and the general sentiment that astronomical observations belong to everyone.

Archival data have been shown[21] to reinforce the legacy value of a telescope project, with the impact increasing with time as the archive grows in size, in terms of number of datasets and size of the data. ALMA tracks how many published studies incorporate data that were requested by one of the authors of a given study, as well as studies in which teams used data only accessible via the public archive. About a quarter of the publications based on ALMA data use at least some archival data. By late 2019, the ASA already hosted more than 1 PB (peta-Byte, or one thousand tera-Bytes) of data.

Publication Statistics

The impact, breadth, and depth of astronomical observations are not trivial to quantify. The main numerical assessment of how prolific a telescope facility has become is the number of scientific publications (peer-reviewed journal articles) that utilize the data from that facility over time. Figure 10.7 shows how the number of ALMA publications has grown since the first early science observing, amounting to over 400 publications per year – more than one per day – after a little less than 10 years of observations.

ALMA is an international project designed to serve the world astronomical community. Figure 10.8 shows the diverse countries to which authors of ALMA research articles are affiliated; more than three dozen countries are represented among authors.

Large Programs

According to ALMA documentation,[22] Large Programs are a subset of observations that "*should address strategic scientific issues that will lead to a major advance or breakthrough in the field.*" Large Programs are those requesting observing time greater than a defined threshold – currently more than 50 hours with the 12 m array or 150 hours with the ACA. Observatories often allocate some fraction of their observing time to this kind of project, seeking results not feasible with one or more smaller projects. Large Programs were first advertised and accepted in 2016, with ASPECS and DSHARP forming the first class.

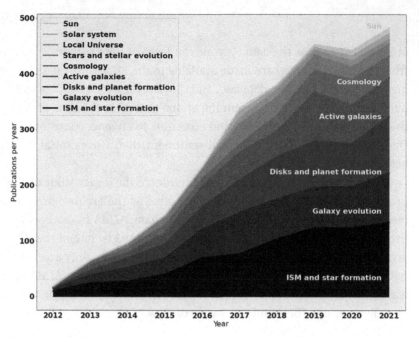

Figure 10.7 Publications in refereed peer-reviewed journals in the first 10 years of ALMA operations, according to topic. Courtesy of F. Stoehr, reproduced by permission.

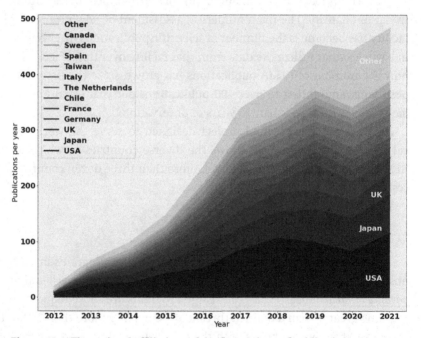

Figure 10.8 The national affiliations of the first authors of publications using ALMA data, during the first 10 years of ALMA operations. The "other" category includes at least two dozen countries. 2,661 publications were reported through the end of 2021. Courtesy of F. Stoehr, reproduced by permission.

A unique aspect of the Large Programs that sets them apart from the other observations is that the team commits to providing high-level data products, free and publicly available, generally via a webpage. These data products are somewhat loosely defined, since the individual programs will naturally necessitate different types of analysis, depending on their scientific objectives. Some examples of the data products are the images that have been generated using different specifications to bring out different levels of detail, along with the scripts used to generate the images. Other examples of data product contributions include spectral energy distributions of observed sources, emission radial profiles, spectra, and source catalogs, among others. The intention is to facilitate sharing of data that enables follow-up or complementary science based on the same observations. These will ensure additional legacy value for the programs, as the astronomical community can access these data products for follow-up studies and analyses.

Very Long Baseline Interferometry

A particularly newsworthy result from a network of telescopes around the world, including ALMA, was the first image of a black hole. A series of six papers[23] were published in a special issue of *The Astrophysical Journal Letters* in 2019 to announce the breakthrough, and reveal the image of the black hole at the center of the galaxy Messier 87, in the cluster of galaxies known as Virgo.

ALMA was certainly not alone in this discovery. Instead, it contributed data along with seven other telescopes: the Atacama Pathfinder Experiment (APEX) in Chile, the IRAM 30 Meter Telescope in Spain, the JCMT in Hawaii, the LMT in Mexico, the SMA in Hawaii, the SMT in Arizona, and the South Pole Telescope. These facilities formed the Event Horizon Telescope (EHT), an Earth-sized interferometer that uses the technique known as Very Long Baseline Interferometry (VLBI). VLBI allows the EHT to achieve an angular resolution of 20 micro-arcseconds, enough to read a newspaper in New York from a sidewalk café in Paris. The EHT team detected light at a wavelength of 1.3 mm and provided direct visual evidence of a supermassive black hole by imaging its shadow, shown in Figure 10.9. Particularly noteworthy, technically speaking, is the additional sensitivity and baselines provided by the Chilean facilities ALMA and APEX, in order to recover the clearly ring-like structure in the image.

Since then, the origin of the jet emerging from the Centaurus A black hole[24] has been imaged with the EHT, and, equally exciting, the detection of polarized light from M87 in 2021. One of the primary goals of EHT was to obtain an image of Sagittarius A* (Sgr A*, pronounced "A-star"), the supermassive black hole at the center of our own Milky Way galaxy. On 12 May 2022, the EHT team presented the image of the Sgr A* black hole that they had obtained. This image was challenging to capture because the black hole spins so rapidly. The EHT Collaboration has been awarded numerous accolades, among them the 2020 Einstein Medal and the 2020 Breakthrough Prize in Fundamental Physics.

Future Prospects

Because technology advances rapidly, ALMA was designed with development in mind. The goal is to keep ALMA at the state-of-the-art for decades to come. Scientific research often leads to more questions than answers. In the case of telescopes, the most advanced technology enables observations that are deeper and more efficient, enabling new discoveries. These additional capabilities are facilitated through the ALMA Development Program, supported by the ALMA partners in addition to the funds they allocate for the annual operation budget. The principles of the development program were restated by the ALMA Board most recently at their 22 April 2020 meeting.[25]

The development program includes two categories: *studies* and *projects*. Studies, being smaller in scope, are administered by the different regions separately. They generally explore concepts that may lead to larger-scale longer-term developments that may be of interest to ALMA. Projects are larger-scale initiatives that provide specific deliverables, either hardware or software, for incorporation into ALMA.

The regions have adopted different strategies and frameworks for encouraging and supporting their development programs.[26] East Asia (EA) solicits

Figure 10.9 Images of the M87 black hole taken with the Event Horizon Telescope. Left: The image from the full EHT array, including data from two telescopes in Chile, APEX and ALMA. ALMA's large collecting area in the Southern Hemisphere added a number of baselines with high sensitivity. Center: How the image would look without the data from the telescopes in Chile. Right: The distribution of telescopes around the globe that contributed to the image. Credit: (left, center) EHT Collaboration; (right) Akiyama et al. (2019); CC BY 3.0.

community feedback at annual EA development workshops, as well as from the East Asia Science Advisory Committee. NAOJ subsequently collaborates with its partner institutes in Taiwan (ASIAA) and Korea (KASI) to lead the development activities. Notable outcomes of the EA development projects are the Band 1 receivers, and the spectrometer for the Total Power Array of the Atacama Compact Array.

The European development program is administered by ESO, which works with ESO member states to solicit and approve about 5–10 studies on three-year cycles. A workshop is generally organized around the time of the call for proposals for development studies. Development programs led by ESO include Integrated Alarm System software for use by the ALMA telescope operators; and an initiative known as Additional Representative Images for Legacy (ARI-L), which reprocesses data products from early ALMA cycles, for ingestion into the ASA. Additionally, the Band 2 receivers were a European development program, with contributions from institutes worldwide, including many across Europe, Japan, Chile, and the United States.

Since 2011 and until the time this book was written, nearly fifty study proposals and a dozen project proposals from the North American community have been primarily funded by the development program in NA. The call for proposals of studies is issued annually, and funding can be requested up to about $250,000, to be used in a timeframe of one year. For the NA call, an independent panel is convened with consent of the NSF, and then an approved ranked list of proposals is incorporated into a recommendation to the NA executive, which has funding authority and responsibility for executing the studies' plans. Several notable development projects led by North American community members include the outfitting of antennas with equipment needed to enable ALMA to become part of the EHT, and several receiver developments, including an upgrade of Band 6, as well as a next generation correlator design. Software tool developments have also been made possible by North American development projects, including the ALMA Data Mining Toolkit (ADMiT) and the Cube Analysis and Rendering Tool for Astronomy (CARTA) used for data visualization.

ALMA2030 is the vision of the ALMA Observatory for the decade of the 2030s. The authors of the ALMA Development Roadmap[27] expressed the community sentiment by stating, "*The three level-one science goals of the ALMA baseline project have been essentially achieved in the first five years of ALMA operations.*" The time had come to propose three new fundamental science goals to explore in the coming decade, and they were laid out with the common theme of origins – of galaxies, chemical complexity, and planets. While publications and the ASA contain hints and tantalizing evidence about these themes, astronomers will need additional capabilities from ALMA in order to search for more conclusive answers. Areas of study include detecting key elements in the first galaxies, tracing organic molecular formation throughout star and planet formation, and ultimately imaging regions where young Earths likely form and where origins of new life are probable.

The top priorities for the 2030 decade are new receivers capable of operating over broader bandwidths, along with an upgrade of the associated electronics, and the ability to process much more data. Additionally, facilitating more efficient data mining of the ALMA archive will lead to greater scientific return. Specifically, the Development Roadmap mentioned the following possibilities: extended baselines by factors of 2 to 3, in order to image nearby terrestrial planet-forming zones; faster wide-field mapping capabilities; increase the number of 12 m antennas; and incorporate a large single dish submillimeter telescope with at least a 25 m diameter. The latter option would enable deep, multi-wavelength images of the sky.

An international partnership is working to design, build, and install a new ALMA correlator, in a project called the Wideband Sensitivity Upgrade (WSU), with the aim of increasing the bandwidth and sensitivity needed by all three science goals of the Development Roadmap. The WSU aims to more than double the bandwidth of the system, meaning that at least twice as much signal information can be recorded in the same amount of time. Also, the spectral resolution, or the fine tuning of the system needed to identify specific molecular emission lines, will be enhanced. Until now, astronomers can be forced to choose between high spectral resolution or broadband recording; the new system should allow the best of both simultaneously.

Conclusion

The journey from first ideas to the reality of ALMA was long and arduous. It stretched 31 years from Owen's suggestion in 1982 that NRAO should build a "millimeter VLA" to the inauguration of ALMA in 2013. From Booth's 1991 proposal to the SEST Users Committee to build a millimeter array in Chile and Ishiguro's 1987 plan for a millimeter array in Japan, there are comparable lengthy time spans. Many, possibly most, of ALMA's users today were barely even born when these initial concepts were set forth. The difficulty of the journey lay in the extensive scope of ALMA and the correspondingly sheer amount of work that was required: refining the scientific goals, designing components, testing prototypes, negotiating contracts and agreements, finding and acquiring a site, assembling the array, and establishing the science centers that would assist users in proposing projects and analyzing data. The work was not resented. Rather, the participants enjoyed and took satisfaction in their efforts, like mountain climbers scaling a peak, focused on a common goal.

The journey was replete with potential hazards. The world of astronomy is densely populated by committees. Observatories appoint committees to hear the wishes of their users and to review their operations. The funding agencies have committees to review and advise on their programs. Community-wide panels set out the priorities for the future. Special situations can arise that call for the appointment of *ad hoc* committees. The drumbeat of reviews was an enormous burden on ALMA, but one that was inevitable. The hazard in every review is the potential for a poor report. Ringing endorsements are necessary from the full array of stakeholders. Tepid enthusiasm will not do. The stakes vary in their consequences. They can be less critical for internal committees but life and death for a review prompted by a crisis. The cost overrun that led to a total reset for the ALMA budget is an example of the latter. ALMA was saved after that review by the commitment to the common goal of transformational science at NSF and ESO.

Circumstances played a role in the success of ALMA, initial disappointments leading to opportunity. The failure of the 25 Meter Telescope project led to a focus in the United States on interferometry and the MMA. The NSF did not immediately fund the MMA, despite outstanding reviews. Had it done so, the United States would now have an array on an inferior site, inadequate to address today's scientific questions. The threat to SEST operations at ESO sparked the drive for the LSA. The lucky detection of CO at $z = 2.3$ boosted the cause of the MMA and led to the LSA's huge collecting area. The timing of the merger of the MMA and LSA was ideal, bridging the budget gap between the VLT and ESO's next generation optical telescope. Only in retrospect do these examples and others show how truly fortunate ALMA was in avoiding dead ends and seizing opportunities.

ALMA triumphed over seemingly intractable problems and cultural differences, but the true triumph of ALMA lies in the scientific discoveries it has enabled. The image of the disk surrounding the young star HL Tauri made the entire journey seem worthwhile. One could almost hear the gasp of amazement and relief from astronomers around the world when the image was released. HL Tauri had long been studied via the infrared emission from its circumstellar disk. ALMA with its sensitivity from its huge collecting area and resolution from its long baselines revealed in stunning detail the structure of the disk – rings of emission from the dust in the disk were separated by gaps where planets were forming. It was the next step toward seeing planets as they form. The cost of building and operating ALMA was a concern of other observatories dependent on NSF funding. At a meeting of a US National Research Council panel, the director of a leading optical/infrared observatory said, "*If I had been shown this image [HL Tau] and told that I could have it for a billion dollars, I would have said – that's a bargain!*" Like the discovery of interstellar CO, the image of HL Tauri was transformational. In its more than 1,000 citations, it is notable that 83 are from PhD theses of young astronomers who have been influenced by this groundbreaking discovery.

ALMA's capabilities are applicable by design to a wide range of research areas, from the Sun to stars, the Galaxy, nearby and distant galaxies. ALMA has transformed the study of the chemistry of interstellar molecular clouds in the Milky Way Galaxy and other galaxies near and far. Observations with ALMA have shed light on solar phenomena, the formation of planets, and the evolution of galaxies in the early Universe. ALMA is a critical element in the EHT, a global-sized very long baseline interferometer. The large collecting area of ALMA has enabled the imaging of the super massive black hole in the galaxy M87 and the one in the center of our own Milky Way. At the time of the writing of this book, ALMA was producing about one scientific article per day based on new observations or those in the ALMA Archive. It would prove to be as productive as the Hubble Space Telescope. With its annual investments in improvements and if the plans for ALMA2030 are realized, ALMA will continue to provide its powerful radio wavelength view of the Universe for very many years. Were ALMA's promises fulfilled? The answer is a resounding YES!

Notes

1 *The ALMA Development Roadmap* can be found at: https://almaobservatory.org/wp-content/ uploads/2018/07/20180712-alma-development-roadmap.pdf.

2 *A Science-Driven Vision for ALMA* in the 2030s (Brogan et al., 2019).

3 The primary paper and survey description for the ASPECS large program is by Walter et al. (2016), and the team's website, with additional publications, a multi-wavelength interactive view, and access to data can be found at: http://aspecs.danielaleitner.de/.

4 Walter to Vanden Bout, private communication.

5 Making an image of HL Tau was the aim of a high-resolution test project involving a team of scientists. The image and analysis were reported by the ALMA Partnership (Brogan et al. 2015).

6 The images were published in the first article of the Astrophysical Journal Letters focus issue (Andrews et al., 2018). Images are also available in the data release: https:// almascience.eso.org/almadata/lp/DSHARP.

7 Herrera's graduate studies (Herrera et al., 2012) were supported by a grant through a partnership between Chile's CONICYT and France's CNRS. Diverse international collaborations not only funded the construction and operations of ALMA, but also the subsequent scientific research.

8 Bae et al. (2022) published *Molecules with ALMA at Planet-forming Scales (MAPS): A Circumplanetary Disk Candidate in Molecular-line Emission in the AS 209 Disk Show Affiliations*, a paper that incorporated data from the MAPS Large Program. Viviana Guzman, who provided a short perspective for this chapter, is a co-author and collaborator.

9 The source IRAS 16293-2422 was originally observed with ALMA in Cycle 1, and the first results were published by Jorgensen et al. (2016) as part of the ALMA Protostellar Interferometric Line Survey (PILS). The same campaign led to at least 29 publications as of 2021. See also Martín-Doménech et al. (2017) and Ligterink et al. (2017).

10 First teams to publish the ALMA Science Verification observations of the Sun were: Bastian et al. (2017), Iwai et al. (2017), Loukitcheva et al. (2017), and Shimojo et al. (2017). All three ALMA partners are represented among the authors of these papers, which appear to have stemmed from coordinated, collaborative efforts. The science verification data are available at http://almascience.org/alma-data/science-verification.

11 Two important, explanatory papers related to Solar observations with ALMA came from Loukitcheva (2019) and Wedemeyer et al. (2020).

12 The research paper by Rizzo et al. (2020) was the result of an effort of seven researchers based in Germany and the Netherlands. Such joint efforts are typical of many ALMA projects.

13 Takafumi Tsukui and PhD supervisor Satoru Iguchi published their results in (Tsukui and Iguchi, 2021).

14 The discovery (Fudamoto et al., 2021) by Yoshinobu Fudamoto from Waseda University and NAOJ came when unexpected light emissions were detected from a region of space that was presumed to be empty and dark.

15 To date, such "open skies" proposals have not exceeded 5% of the total ALMA observing time for any given cycle. The excess time is to be allocated to the North American share, per current US government policy associated with funding from the National Science Foundation.

16 Each proposing team designates a reviewer from among their team. Also, a reviewer without a PhD has to indicate a "mentor."

17 A full overview is given by Donovan Meyer, J. et al. (2022).

18 Carpenter et al. (2022) give an overview and update on the ALMA proposal review process, with a focus on systematics between Cycles 0 and 8. They report some biases in earlier cycles related to the relative prominence of the Principal Investigator (PI), a so-called "prestige bias." The report does not report any findings of significant gender biases. They also looked at systematics related to regional affiliations of proposal authors.

19 A multi-disciplinary research team of experts from Canada, Switzerland, and Chile met at ALMA in April 2016 to examine observatory staff and study the effect of oxygen deficiency, a medical condition known as hypoxia, when working at high-altitude (Pun et al., 2018).

20 Data from the ALMA telescope needs to be processed before delivery to the respective astronomers. This is known as "data reduction." A software pipeline has been developed for this purpose, and the vast majority of data from the telescope pass through this pipeline processing. Data reduction tasks were initially distributed among the ARCs (Schnee et al., 2014), and in recent cycles, have been primarily the responsibility of a data reduction team at JAO. By Cycle 8, nearly 90% of data were processed at the JAO, compared with about 50% in Cycle 5, and in earlier cycles, the majority of data were processed at the ARCs. The ARCs complete the process with a final review and quality assessment before delivering data to the community.

21 See Peek (2017). www.scientificamerican.com/article/how-old-observations-are-building-hubbles-legacy/.

22 For a detailed description of Large Programs, see https://almascience/alma-data/lp.

23 The ALMA press release can be found at: www.almaobservatory.org/en/press-releases/astronomers-capture-first-image-of-a-black-hole/. The *Astrophysical Journal Letters* organized a special issue with the title "*Focus on the First Event Horizon Telescope (EHT) Results,*" where Shep Doeleman, on behalf of the EHT Collaboration opened with "*We report the first image of a black hole.*" The article proceeds to explain the ambitious observing campaign over the past decade, and the extensive coordination that led to the subsequent publications. After the first (Akiyama, K., et al., 2019) of six papers were published in 2019, two additional papers in 2021 expanded the study to the magnetic fields of M87. See, for example, Akiyama, K. et al. (2021).

24 Janssen et al. (2021) explain that the SMBH in Centaurus A bridges the gap in mass and accretion rate between that of Messier 87 and that of our Galactic Center.

25 *Principles of the ALMA Development Program* can be found at: https://science.nrao.edu/facilities/alma/science_sustainability/Principles_of_the_ALMA_Development_Program.pdf.

26 Summaries of the development programs of the three ALMA regions are given in *The ALMA Development Program: Roadmap to 2030* by Carpenter, J. et al. It can be found at: https://science.nrao.edu/facilities/alma/science_sustainability/URSI.pdf.

27 The *ALMA Development Roadmap*, J. Carpenter et al. (The ALMA Development Working Group) can be found at: www.almaobservatory.org/wp-content/uploads/2018/07/20180712-alma-development-roadmap.pdf.

Appendix A

Radio Astronomy

Radio Waves

This is a book in which radio telescopes play a central role. While we think it is likely that most readers of our book are familiar with that subject, we would also guess that a minority of our audience (and most of the general population) do not know all that much about radio telescopes and the scientific revelations they have produced. Our aim here is to remedy this situation for our non-astronomer readers as briefly and non-technically as possible.

All astronomical telescopes share the same basic functional structure. Radio telescopes image celestial phenomena just as optical telescopes, but they do so in radio waves rather than visible light. Humans are simply not equipped to see radio waves. You can look through even a small telescope and see the mountains and craters of the Moon, but without complicated equipment, it is not so easy to detect even the strongest radio waves that reach Earth from interstellar space. Nonetheless, the underlying commonality of all telescopes is the result of the fact that the light by which we see turns out to be part of a fundamental and far grander phenomenon that is largely invisible to the eye, namely electromagnetic waves.

Visible light is an electromagnetic wave. The different colors of the light seen in a rainbow, are part of the same phenomenon, simply made up of light of slightly different wavelengths. More important, theory does not forbid electromagnetic waves of radically different wavelengths, ranging from the ultra-short (what we now call x-rays and gamma rays) to the comparatively long (what we now call the radio regime). Figure A.1 illustrates the bands of the electromagnetic spectrum.

211

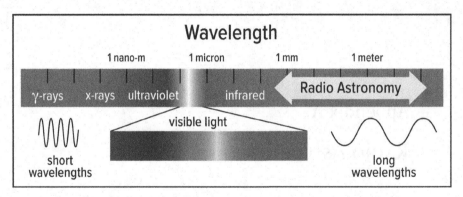

Figure A.1 The electromagnetic spectrum from gamma rays to radio waves. Visible light occupies only a narrow slice compared with the radio band. Credit: Jeff Hellerman; NRAO/AUI/NSF, CC BY 3.0.

The characteristics of any wave are its wavelength (distance between successive wave peaks), its frequency (the number of waves passing a fixed point per second), and its speed. By definition, wavelength times frequency is equal to speed. The speed (in a vacuum) of electromagnetic waves, produced by accelerated charges, is the speed of light. The modern term "radio" usually includes all electromagnetic waves longer than the infrared. This is actually a huge span of wavelengths, running from about 100 km (approximately 60 miles) to around 1,000 millionths of a meter (1,000 microns, or about 1/25th of an inch). Obviously, dealing with such a large range of numbers can become very cumbersome, and for convenience, it is usual for scientists and engineers to subdivide the radio regime into smaller domains. Electromagnetic signals with wavelengths shorter than about 10 cm are commonly called *microwaves*, waves shorter than 1 cm are called *millimeter waves*, and waves shorter than 1 mm *submillimeter waves*.

Radio telescopes observe four types of cosmic radiation: *synchrotron*, *bremsstrahlung*, *thermal*, and *spectral line*. Figure A.2 illustrates the first of these mechanisms. Synchrotron radiation is produced by free electrons following spiral paths around magnetic field lines. The electrons are constantly accelerated while following these paths and they emit electromagnetic radiation which is strongest at low frequencies in the radio band. Radio astronomy began with the observation of synchrotron radiation. It was the discovery of such radiation from the center of our Galaxy by Karl Jansky in 1933 that marks the birth of radio astronomy. Figure A.3 is an image of synchrotron emission captured by the Very Large Array (VLA) of the radio galaxy Hercules A.

Bremsstrahlung radiation is emitted from a plasma of ionized (largely) hydrogen atoms. The electrons are deflected in collisions with the protons, emitting radio waves. The process is illustrated in Figure A.4. The plasma is

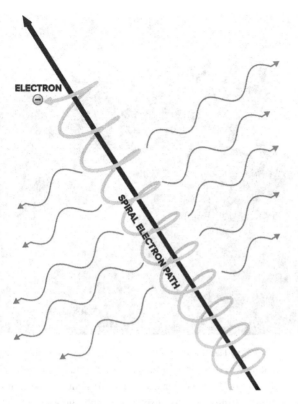

Figure A.2 Synchrotron radiation produced by an electron spiraling in a magnetic field. Credit: Bill Saxton; NRAO/AUI/NSF, CC BY 3.0.

created in the interstellar medium by nearby hot, luminous stars. The Orion Nebula is a prominent example in our Milky Way Galaxy, shown in Figure A.5.

Thermal radiation is produced by the random motions of electrons in a warm object; that is, any object with a temperature above absolute zero. All the room temperature objects in our daily environment, including our bodies, radiate in the infrared. At the higher temperatures found, for example, in flames, the radiation is visible; we see the glow of heated objects. Stars have temperatures that cause them to radiate in the visible light band and in the adjoining infrared and ultraviolet bands. The light thus recorded by an optical telescope is almost always starlight, and objects in the Solar System are seen in reflected light from the Sun. Images of our Galaxy and external galaxies are composed of a myriad of stars, as Galileo found when he turned his telescope to the Milky Way. In contrast, the cosmic background radiation, cool in temperature, is strongest at millimeter wavelengths. It is thermal radiation from the Big Bang. Thermal radiation from the dust in protoplanetary disks is seen in the images made by ALMA (see Chapter 10, Figures 10.1 and 10.2). ALMA, the

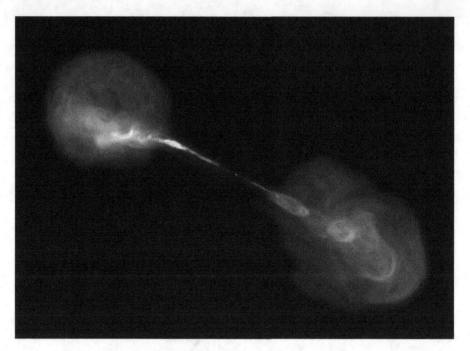

Figure A.3 The radio galaxy Hercules A as imaged by the VLA. The narrow jets emanate from a black hole in the center of the galaxy and end in the large lobes. The synchrotron emission of the radio image reveals a structure not seen in visible light; the optical emission from the galaxy itself would only show a small dot in the center. Credit: Bill Saxton, Bill Cotton, and Rick Perley; NRAO/AUI/NSF, CC BY 3.0.

BREMSSTRAHLUNG RADIATION

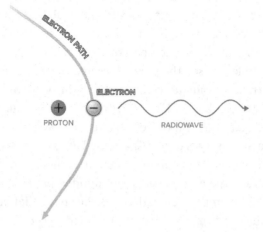

Figure A.4 An electron deflected by an ionized atom radiates bremsstrahlung radiation. Credit: Bill Saxton; NRAO/AUI/NSF, CC BY 3.0.

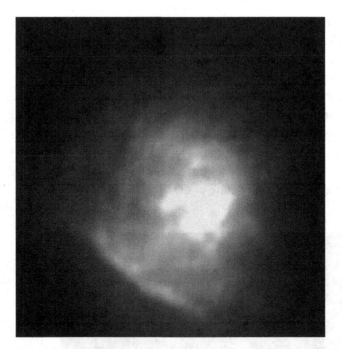

Figure A.5 The Orion Nebula imaged in bremsstrahlung radiation by the VLA. Such radiation implies a plasma of ionized atoms in this star-forming region. Credit: Debra Shepherd, Ron Maddalena, and Joe McMullin; NRAO/AUI/NSF, CC BY 3.0.

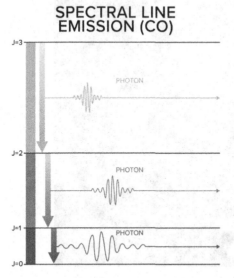

Figure A.6 The three lowest frequency transitions between the quantized energy levels of CO: $J = 1$–0 at 115 GHz, $J = 2$–1 at 230 GHz, and $J = 3$–2 at 345 GHz. These frequencies of CO are the workhorse observational frequencies for ALMA, among the thousands of frequencies from the two hundred plus interstellar molecules. Credit: Bill Saxton; NRAO/AUI/NSF, CC BY 3.0.

subject of this book, is an instrument exquisitely designed for observations of thermal radiation to study the "cold universe."

Spectral line emission occurs because the energy of atoms and molecules is confined to particular states or energy levels. For example, the tumbling motion of the carbon monoxide (CO) molecule cannot proceed at an arbitrary rate but must occur in quantum states labeled by a number that indicates their spin rate. Figure A.6 illustrates the source of three lowest frequency spectral lines of CO, namely transitions from the $J = 1$, 2, and 3 energy states to the next lowest energy state. Figure A.7 is an image[1] of the galaxy M51 in the radiation from the $J = 1$ to $J = 0$ quantum state of CO.

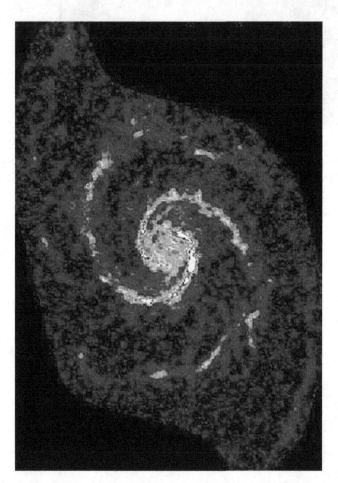

Figure A.7 M51, the Whirlpool Galaxy, imaged in the radiation from the lowest frequency spectral line of carbon monoxide, CO, at 115 GHz. The image combines data obtained by CARMA and the Nobeyama 45 m Telescope. Credit: Koda et al. (2009); ©AAS, reproduced by permission.

Telescopes

In 1609, Galileo Galilei, who held the Chair of Mathematics at the University of Padua and who was helping to move the science of physics out of the swamp of Greek philosophy and into the realm of verifiable natural law, fabricated a small telescope. Galileo did not invent the telescope.[2] However, he was, as far as is known, the first human being to turn a telescope on the night sky, explore what he found in a systematic way, and analyze and publish his findings. Galileo's first telescope only magnified by a factor of three, but just a few months later he had built a 20-power instrument. Observations with it revealed the Moon to be a world unto itself, with mountains, craters, and what seemed to be (but were not) seas and oceans, and led to the discovery of the four largest moons of Jupiter; other observations showed that the planet Venus exhibits phases like those of the Moon, and that what appeared to the naked eye to be diffuse nebular patches of light in the Milky Way could be at least partially resolved into stars using his telescope. Galileo continued to improve his designs, and by 1611 further observations showed the Sun was not the perfect orb presumed by philosophers; instead, it had a surface harboring small, dark spots that revealed a slow rotation about its axis.

For the next 320 years, every astronomical telescope built on Earth was essentially an elaboration of Galileo's first instrument, designed with one fundamental goal: to extend the reach of human vision. Beginning in the early twentieth century, astronomers began using new technologies to build instruments aimed at portions of the electromagnetic spectrum that lay beyond the narrow range of wavelengths directly visible to the human eye, and by the latter part of the twentieth century, astronomers had added gamma-ray, x-ray, ultraviolet, infrared, and radio telescopes to their kit.

Nevertheless, most modern radio telescopes share a common basic architecture: a large collecting element to gather and focus faint light from the sky. This is followed by a "receiving system" which detects and amplifies the light, digitizing and recording an image of the observations. This latter process is directly comparable to the way the lens, electronic sensor, and memory card of modern digital cameras operate to produce a digital image.

The generic structure of a typical radio telescope is quite similar to that of optical reflecting telescopes. Notably, there is a signal collecting element – usually a reflecting surface – that is functionally identical to the main mirror and subsidiary light-handling elements of an optical telescope. Radio waves that are captured by the reflector are then focused to go through a sequence of electronic systems that detect, amplify, digitize and finally record the captured signals, just as in the case of modern optical telescopes and digital cameras. Radio astronomers still use the descriptive term "receiver" for the electronics responsible for the initial stage of signal detection. An important common feature of most optical and radio telescopes arises because both types of instruments must have the ability to continuously follow astronomical sources across the sky in order to compensate for their apparent motion caused by the Earth's rotation. Figure A.8 shows the functional outline of a single-dish radio telescope.

While single-dish radio telescopes typically have a close one-to-one correspondence with their optical counterparts functionally, sometimes more specialized radio telescope designs are required to overcome technical or physical performance limitations. Invariably, such designs will involve tradeoffs. We discuss one of these next.

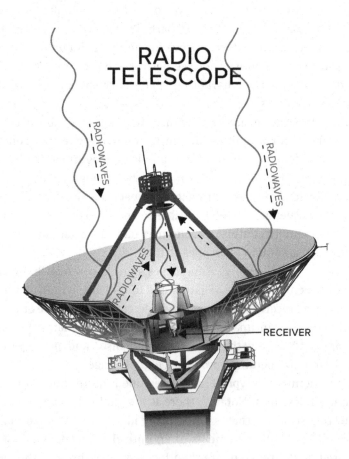

Figure A.8 Conceptual diagram of a single-dish radio telescope. Credit: Bill Saxton; NRAO/AUI/NSF, CC BY 3.0.

Radio Interferometers

A particularly acute performance limitation of single-dish radio tele-scopes is their relatively poor resolution. The term *resolution* here simply means the detail that can be seen in an image. It is measured as the angle subtended by two barely discernable objects as seen by the observer. The resolution of any telescope that is not limited by atmospheric distortion is proportional to the wavelength of the electromagnetic waves it is designed to detect divided by the size of its main reflector. Since radio waves are defined by having enormously larger wavelengths than visible light waves, this means that a radio telescope must be enormously larger than any given optical telescope if it is to provide an image of comparable detail. Consequently, a single-dish radio telescope operat-ing at a wavelength of 1 mm would have to be nearly 3 km in diameter if one wanted to produce radio images with the same level of detail as the Hubble Space Telescope.

Building a steerable, rigid, radio telescope dish 3 km in diameter is a struc-turally hopeless task. An antenna that size would have to be supported at its center, which would therefore have to sit on a pedestal or other support struc-ture around 1.5 km in height. This is about twice the height of the Burj Khalifa, the tallest building yet built by the human race. More to the point, the largest steerable radio telescope in the world has a main dish around 100 m in diame-ter, close to the limiting size possible.[3]

It is this frustrating situation that led radio astronomers to develop instru-ments called *imaging interferometers*, which can provide the resolution of much larger radio telescopes by using many smaller radio antennas to at least partially fill the "footprint" of the larger antenna they are unable to build. Superficially, an imaging interferometer looks like an array of similar radio telescopes that observe the same source at the same time. This appear-ance masks a significant increase in complexity. In an interferometer, the individual telescope outputs are connected to a highly specialized computer, called a *correlator*, which combines the data streams to form an image having a resolution comparable to what would be produced by an antenna whose diameter is equal to the largest separation of the various small antennas in the array.

A price is paid for the gain in resolution. An imaging interferometer is in some ways a poor astronomer's giant telescope, that is, one missing most of its reflecting surface. It is a "sparsely filled" aperture. While the notional giant antenna would catch all the radio waves that fall on its reflecting surface, a sparsely filled aperture of the same size can only collect signals that happen to fall on the parts of the giant antenna's footprint that are occupied by smaller

Figure A.9 Ten of the 27 antennas that make up the Very Large Array in New Mexico. Credit: NRAO/AUI/NSF, CC BY 3.0.

antennas. Thus, an imaging interferometer cannot match the sensitivity of the ideal giant telescope. There is also an important technical issue associated with the images produced by sparse arrays. Due to missing reflector elements relative to the ideal fully filled surface, there is irretrievable missing information needed to fully reconstruct radio images. Put another way, the images produced by radio interferometers are *estimates* of what a giant full aperture telescope would yield, and they require substantial off-line computer processing to attain their full potential.

NRAO's VLA, built during the 1970s and dedicated in 1980, was the first large imaging interferometer. (Large interferometers are called arrays.) The VLA is shown in Figure A.9. Although interferometers are much more complex than single-dish telescopes, they have become indispensable instruments to modern radio astronomy. ALMA is an imaging radio interferometer, the product of a long history in the development of radio telescopes.

Misconceptions

Radio telescopes have nothing whatsoever to do with radio and television sets or their broadcasts, although all involve radio waves. But this misconception is remarkably common. A resident of Green Bank, West Virginia, once complained to the Observatory that the radio astronomers were watching her through her television set, which would have been a stunning technical achievement were it even possible. Another common misconception is that radio astronomers only listen for messages from extra-terrestrial civilizations. John Kraus, a pioneer in radio astronomy, implied just that when he named the telescope he used to search for such signals the "Big Ear." The Search for Extra-Terrestrial Intelligence (SETI) does use radio telescopes to search for artificial signals, a sign of non-terrestrial life. In 1960, the first such search was conducted at NRAO's Green Bank site using the Tatel Telescope, an 85 ft diameter antenna. The search was called Project Ozma[4] and was conducted by NRAO staff members, led by Frank Drake. The movie *Contact*, based on the novel of the same name by Carl Sagan, showed the star Jodie Foster listening with headphones connected to the VLA. Radio astronomers today do not use headphones, nor do they "listen." The popularity of the movie reinforced this notion in the general public's concept of radio astronomy. But SETI is only a small aspect of radio astronomical activity. The vast majority of radio astronomers are studying the natural emission in radio waves from celestial objects, not searching for artificial emission transmitted by a distant civilization.

Appearances can lead to misconceptions. Satellite television and radio systems may employ ground stations that include one or more large dish-type antennas for downloading content, but these are not radio telescopes. Likewise, space agencies such as NASA and sister organizations elsewhere in the world that track and operate deep space missions send commands to their spacecraft and download scientific data using large antennas that resemble radio telescopes. Their mission and purpose is not radio astronomy. Not every antenna is a radio telescope.

Finally, some radio telescopes look nothing like a dish. This is particularly true at long radio wavelengths. A classic example is the antenna, shown in Figure A.10, that Karl Jansky used in 1932 to make his discovery of radio waves from the center of the Milky Way. This was the first detection of radio waves from a non-terrestrial source, the beginning of radio astronomy.

In summary, beginning with Jansky, radio astronomy has revealed an entirely new view of the Universe, one rich in discoveries.[5] ALMA is the most recent example of the power of radio telescopes to advance our understanding of cosmic phenomena. We can look forward to the era of the New Generation Very Large Array (ngVLA) and the Square Kilometer Array Observatory (SKAO), which will accomplish at lower frequencies what ALMA has done at higher frequencies.

Figure A.10 Karl Jansky standing in front of his discovery making radio telescope. Credit: Nokia Corporation and AT&T Archives, reproduced by permission.

Notes

1 Koda et al. (2009).
2 However, the telescope was a fairly recent invention in Newton's time, as documented by two patent applications to the Dutch government that were never granted. By 1609, details of the simple design had been widely disseminated throughout Europe. The development of lenses and the telescope has been described by the Galileo Project. http://galileo.rice.edu/sci/instruments/telescope.html.
3 The US Naval Research Laboratory attempted to build a 600 ft diameter antenna in the late 1950s at a site in Sugar Grove, Virginia, near the NRAO Green Bank site. The project was a fiasco and it was canceled in 1962 by Defense Secretary Robert McNamara. See Kellermann, Bouten, and Brandt (2020) for a description of the project.
4 For a history of SETI, see Drake and Sobel (1992).
5 For an account of radio astronomy discoveries, see Kellerman and Bouton (2023).

Appendix B

Millimeter/Submillimeter Telescopes

US Millimeter Telescopes

Here we briefly introduce and summarize a number of the most impactful single-dish millimeter wavelength telescopes constructed in the late twentieth century. These built the legacy of millimeter wave astronomy that led to ALMA.

University of Texas Millimeter Wave Observatory – Observations began with the 4.9 m diameter telescope of the Electrical Engineering Research Laboratory of the University of Texas in June of 1963, at its off-campus site at the J.J. Pickle Research Center in the northern outskirts of Austin, Texas. After measuring the brightness at 35, 70, and 94 GHz of a number of planets, the antenna was moved to Mt. Locke, the site of McDonald Observatory in the Davis Mountains of West Texas, where it was housed in an astrodome, shown in Figure B.1. In 1971, the antenna was "discovered" by Pat Thaddeus and one of the authors (Vanden Bout), who were observing on the McDonald 107 Inch Telescope. The antenna was idle. By the fall of 1972, a partnership began using the Millimeter Wave Observatory (MWO) to exploit the potential of the recently discovered interstellar CO. Bell Labs contributed the receivers, the Goddard Institute of Space Studies (GISS) purchased the klystron oscillators required for the receiver system, and Harvard College Observatory built the spectrometer or "backend." The observing time was shared equally among the four partners. The MWO had a large number of observers from the United States and abroad, and MWO data have been included in at least 23 doctoral dissertations. Research tended to be concentrated on mapping individual CO sources in the Galaxy. Perhaps

Figure B.1 The MWO 4.9 Meter Telescope in its renovated astrodome equipped with an error-correcting subreflector. Credit: McDonald Observatory, University of Texas at Austin, reproduced by permission.

the most significant discovery was of jets of gas known as bipolar outflows that are the by-product of young stars, most importantly, the source known as L1551.[1] The MWO was decommissioned in 1988. The reflecting surface of the 4.9 Meter Telescope is at the campus of the Instituto Nacional de Astrofísica, Óptica y Electrónica in Tonanzintla, near Puebla, Mexico[2]; the astrodome and mount have been moved to Sierra Negra, the site of the Large Millimeter Telescope (LMT).

Columbia University Mini Telescope – In 1974, Thaddeus and his group at GISS installed a 1.2 m diameter antenna in a dome on the roof of the Pupin Physics Building of Columbia University. The relatively small size of the antenna was deliberately chosen, as a smaller telescope with its wide field of view can rapidly construct maps of the CO distribution in the plane of the Milky Way. The "Mini" was highly successful, producing maps of Galactic CO, which led to numerous studies[3] interpreting these maps and features within. Figure B.2

Figure B.2 The Southern Mini in its astrodome at CTIO. Courtesy of Thomas Dame, reproduced by permission.

shows the Southern Mini, installed in 1982 at the Cerro Tololo Inter-American Observatory (CTIO) in Chile, which supplied the complementary data from the southern hemisphere. Observations of Galactic CO with the Southern Mini were conducted by Leo Bronfman of the U. Chile. The first Mini was moved to Harvard College Observatory in 1986 and its operation continues to the present under the leadership of Tom Dame.

Aerospace Corporation Millimeter Telescope – The Aerospace Corporation is a research organization originally supported by the Air Force to conduct defense research. It also had a vigorous program of research in optical and x-ray astronomy through the 1970s when a 4.6 m diameter antenna for millimeter wavelength observing was constructed. The antenna resided on the roof of the Aerospace building just south of the Los Angeles International Airport. It had been built by the Rohr Corporation, shortly before Rohr built the NRAO 36 Foot Telescope. The proximity to the Pacific Ocean made for a less than desirable site for a millimeter telescope but it was adequate for observations at 3 mm wavelength. It made observations of planets, quasars, and the association of CO emission with individual objects in the Galaxy. A report of the activities in 1975 is typical of the observing programs[4] which involved mapping CO emission from Galactic molecular clouds. A significant contribution of the Aerospace telescope was a map of the emission from the two isotopo-logues, ^{13}CO and ^{12}CO, in L134 which showed that the abundance of ^{13}CO with respect to ^{12}CO increased toward the center of the molecular cloud, as was predicted by models of isotopic fractionation in dense clouds. Anneila Sargent commuted from Caltech in Pasadena to make the observations for her PhD dissertation using the Aerospace antenna. The staff of the Aerospace Telescope included William Wilson, Eugene Epstein, and one of the authors (Dickman), who was in charge of operations.

Five College Radio Astronomy Observatory – The Five College Radio Astronomy Observatory (FCRAO) was organized in 1970 as a partnership between Hampshire, Amherst, Smith, Mount Holyoke Colleges, and the University of Massachusetts, Amherst. The first research project was the study of pulsars using a cluster of four "mini-Arecibo" telescopes, consisting of reflecting mesh draped between utility poles. The telescope was located in a wooded reserve at Quabbin Reservoir. This work led in time to the discovery by Hulse and Taylor, at the Arecibo Observatory, of a binary pulsar with a millisecond period. The properties of the pulsar could be explained if it emitted gravitational waves. In the meanwhile, research had changed directions at FCRAO, turning to inter-stellar molecular spectroscopy.

In late 1972, ESSCO offered to provide a 14 m diameter antenna, in a protective radome, with a surface of accuracy sufficient for observing at milli-meter wavelengths. The telescope was dedicated in 1976 and reached routine operation in 1978. Dickman was the FCRAO manager for a number of years. The radome is shown in Figure B.3. Although studies were made of multiple molecular species, its *forte* was surveys of CO in our Galaxy, the Milky Way, and in nearby galaxies. The Milky Way surveys were of necessity more limited in extent than those of the Columbia Mini because of the more than 100 times finer resolution in the maps produced, but although they take took longer, more detail was discerned in the final product. The resulting detail allowed for significant new results. For example, it became clear that most of the inter-stellar molecular gas is in huge, dense clouds, called Giant Molecular Clouds (GMCs). The bulk of star formation in our Galaxy, and presumably other gal-axies, occurs in GMCs. FCRAO was also well known for its technical expertise, most notably its array receivers: the 16-element QUARRY and the 32-element SEQUOIA arrays. FCRAO closed in 2006, but its legacy lives on in the careers of many leaders in millimeter astronomy. Its receivers have been repurposed and installed in other telescope projects around the world, most notably, the LMT in Mexico.

Figure B.3 The radome of the FCRAO 14 Meter Telescope at Quabbin Reservoir. Courtesy of Mark Heyer, University of Massachusetts at Amherst, reproduced by permission.

Bell Telephone Laboratories Telescope – Following years of observing at the MWO, the Bell Labs team built their own antenna. It had a unique design that featured an unblocked aperture of 7 m diameter with nothing between the reflecting surface and the sky. The telescope was completed in 1978.[5] Work at the facility included studies of isotopic species of interstellar molecules, observations of individual molecular clouds, and surveys[6] of CO emission, particularly the ^{13}CO isotopologue. The observing was largely done by Bell Labs staff members, but there was also a visiting observer program. The 7 Meter Telescope is shown in Figure B.4.

Millimeter Telescopes Outside the United States

The establishment of millimeter wavelength observatories in the United States was quickly followed abroad. Two large-diameter telescopes were built in the early 1980s which completely outclassed all those in the United States.

Institut de Radio Astronomie Millimétrique – The Institut Radio Astronomie Millimétrique (IRAM) had a somewhat fraught beginning.[7] In the 1970s, discussions among the French, German, and British radio astronomers led to an

Figure B.4 The Bell Labs 7 Meter Telescope with its unblocked aperture. Arno Penzias is in the foreground. (Note: This is not the horn telescope Penzias and Robert Wilson used for their discovery of the cosmic background radiation.) Credit: Nokia Corporation and AT&T Archives, reproduced by permission.

agreement to form the Joint Institute for Millimetre Astronomy. The plan collapsed when the British withdrew. This left the French and German participants to decide what they alone would build, the Germans favoring a large 30 m diameter antenna and the French a millimeter interferometer. When Spain joined the partnership, it became possible to do both. IRAM was founded on 2 August 1979. The 30 Meter Telescope, shown in Figure B.5, was the first of the IRAM facilities to be completed. It is located on Pico Velata near Granada, Spain. Later, an interferometer was completed on the Plateau de Bure, near Grenoble, France. The upgrade of the Plateau de Bure Interferometer, the Northern Extended Millimetre Array (NOEMA) is in operation today, as the premier interferometer for millimeter wavelengths in Europe. IRAM is also a major center for the development of radio astronomy instrumentation and technology.

Figure B.5 The IRAM 30 Meter Telescope on Pico Velata, Spain. Credit: ©IRAM, reproduced by permission.

Nobeyama Radio Observatory – Japanese radio astronomy began in the late 1940s with observations of the Sun by researchers at several institutions. A community-wide forum for the promotion of radio astronomy in Japan was organized in December 1969 called the Japan Radio Astronomy Forum. The first millimeter wave telescope was a 6 m antenna installed in 1970 at the Mitaka site of the Tokyo Astronomical Observatory, which later became the National Astronomical Observatory of Japan (NAOJ). In the early 1970s, a plan was developed to construct a 45 m diameter telescope in combination with a five-element millimeter interferometer. These efforts combined at the Nobeyama Radio Observatory (NRO) under the leadership of Masaki Morimoto. The 45 Meter Telescope was to operate at wavelengths from 3 mm to 21 cm. Construction began in 1978 and was completed in 1982. The NRO 45 Meter Telescope continues operations today. It has a novel acoustic spectrometer of 16,000 channels that has been used to discover a number of interstellar molecules. The NAOJ Advanced Technology Center is a major player in the development of millimeter/submillimeter receivers and instrumentation. The 45 Meter Telescope is shown in Figure B.6.

Figure B.6 The NRO 45 Meter Telescope in its mountain valley site. Credit: ©NAOJ, reproduced by permission.

Onsala Space Observatory – The Onsala Space Observatory (OSO) was founded in 1949 by Olof Rydbeck of Chalmers University. The OSO was dedicated in 1955 and its signature telescope, a 25.6 m diameter antenna, was first operational in 1963. Among the chief programs of research were studies of interstellar CH spectral lines at 9 cm wavelength. These lines were used to study the large-scale structure of the Milky Way. In 1976, a near-twin of the FCRAO millimeter telescope was completed at the OSO. Projects undertaken with this 20 m diameter telescope included a survey of interstellar spectral lines between 72 and 91 GHz emitted from the Orion Nebula and a dust-enshrouded star known as IRC +10216, observations showing that the CO emission from the galaxy M51 follows the spiral arms seen in visible light, and numerous other projects conducted by staff and visiting observers.

Submillimeter Telescopes

One can argue that the history of radio astronomy is a progression from observations at long to ever shorter wavelengths. The field began with

Figure B.7 The SEST telescope at ESO, La Silla, Chile. Credit: ESO, CC BY 4.0.

Karl Jansky's discovery of galactic radiation at a wavelength of 14.6 m. In contrast, interstellar CO was discovered at a wavelength of 2.6 mm. The factor of nearly 5,600 in wavelength between these observations was made possible by advances in receiver electronics. As that progression in technology continued, it was inevitable that astronomers would turn to the submillimeter band, that is, wavelengths shorter than 1 mm. Accordingly, starting in the 1980s and up to the time of the writing of this book, the following eight single-dish submillimeter telescopes were built.

James Clerk Maxwell Telescope In 1983, a partnership between the United Kingdom (UK), the Netherlands, and Canada built the James Clerk Maxwell Telescope (JCMT) on Maunakea. It had a high-accuracy reflecting surface of 15 m diameter capable of submillimeter observations. The main instrument of the JCMT was a pair of receiver arrays, composed of bolometer detectors, called SCUBA. Using SCUBA, observers were able to establish a new class of galaxies called submillimeter galaxies. Today, the telescope has a much more powerful array detector, SCUBA-2, and is operated by a consortium of East Asian countries and UK universities.

Sweden-ESO Submillimetre Telescope – The success of the OSO 20 Meter Telescope's programs led to the construction of the 15 m Sweden-ESO Submillimetre Telescope (SEST) at ESO's La Silla site in Chile. Among its successes was a survey of spectral lines from Sgr B2, a strong molecular line source near the center of the Galaxy, over the frequency range 225–250 GHz. Figure B.7 is a photograph of SEST. It was the second millimeter-wavelength telescope built in Chile; a small 60 cm diameter dish had been installed earlier as a test facility by the NAOJ. SEST was decommissioned in 2003. Plans have been announced for moving the SEST antenna to Namibia, where it would serve as a far-south element of the EHT.

Caltech Submillimeter Observatory – Bob Leighton built a total of seven aluminum honeycomb sandwich reflector antennas of diameter 10.6 m, Dave Woody overseeing the construction and implementation of the last three. Six went into the OVRO Millimeter Array and the one with the best surface accuracy was used at the Caltech Submillimeter Observatory (CSO), shown in Figure B.8. The CSO was located in the saddle area just below the peak of Maunakea in

Figure B.8 The Caltech Submillimeter Telescope in its astrodome on Maunakea. The JCMT dome is in the upper left. Courtesy of Sunil Gowala, reproduced by permission.

Figure B.9 The Submillimeter Telescope (SMT) of the University of Arizona. It is one of three telescopes on Mt. Graham, the others being the Large Binocular Telescope and the Vatican Advanced Technology Telescope. Credit: ESO, CC BY 4.0.

Hawaii. The measured surface accuracy using astronomical sources is about 20 μm. Submillimeter observing requires very low atmospheric water vapor and even on Maunakea this could be challenging. The CSO was only operated at night to minimize thermal deformations of the reflecting surface. An astrodome reduced the effect of wind. CSO operations began in the 1990s and continued until it was decommissioned in 2015. Over that period the detectors and associated analysis software were steadily improved for use in a variety of research programs. Tom Phillips was the long-time director of the CSO and has written a summary[8] of its history and scientific results. Among the more significant are the spectral line surveys, and the discovery that deuterated molecules were much more abundant than anticipated, with even triply deuterated ammonia detectable.[9] On 2 March 2021, the Maunakea Management Board approved a plan to deconstruct the CSO and restore its site in 2022. It has been suggested that the CSO be moved to Chile.

The Submillimeter Telescope – The push to the submillimeter band also led to the construction of the SMT at the University of Arizona. The SMT is a 10 m diameter telescope housed in an enclosure with large doors that open to the sky. Its construction was completed in 1993 following a fractious debate over the possible impact on the endangered Mt. Graham red squirrel, a situation settled by an act of the US Congress. It was originally built in partnership with the MPIfR and at that time called the Heinrich Hertz Telescope. The SMT is shown in Figure B.9, peering out between the massive open doors of its enclosure.

Mt. Fuji Submillimeter Telescope – In 1998, the Japanese, under the leadership of Satoshi Yamamoto, began operation of a submillimeter telescope[10] on the summit of Mt. Fuji, at an elevation of 3,725 m. The antenna had a diameter of 1.2 m and was housed in a rotating dome. It was equipped with receivers operating in the 345, 492, and 810 GHz bands and an acoustic-optical spectrometer of 1,024 channels. The telescope is shown in Figure B.10. The principal observing program of this telescope was a survey of emission from neutral carbon atoms in the interstellar medium of the Milky Way.

Cologne Observatory for Submillimeter Astronomy – In 1985, Gisbert Winnewisser installed a 3 m diameter antenna in the south tower of the Külm Hotel on the Gornergrat peak above the Swiss village of Zermatt. The telescope's

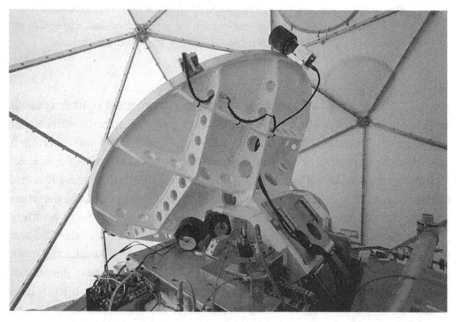

Figure B.10 The Mt. Fuji Telescope in its radome. Courtesy of Yutaro Sekimoto, reproduced by permission.

Figure B.11 The Atacama Submillimeter Telescope Experiment (ASTE) at Pampa la Bola east of the ALMA site on the Llano de Chajnantor. Credit: ©NAOJ, reproduced by permission.

receivers, covering the frequency range of 210–820 GHz, were used to study interstellar molecular lines. In 1985, the telescope was upgraded with a new mount and reflecting surface. In a partnership with the National Astronomical Observatories of China, the telescope was moved to Tibet in 2010 to become the China-Cologne Observatory for Submillimeter Astronomy. Telescope operation resumed in 2013 following the installation of improved instrumentation.

Atacama Submillimeter Telescope Experiment – Over the two-year period 2002–2004, the NAOJ installed a 10 m diameter telescope at Pampa la Bola, near the future site of ALMA. The site is at an elevation of 4,860 m, slightly lower than the ALMA site. The Atacama Submillimeter Telescope Experiment (ASTE) supports spectroscopic observations at frequencies as high as 350 GHz, or a wavelength as short as 0.87 mm. Figure B.11 shows the telescope with a vehicle for size comparison. ASTE has been a very productive facility since it began operation, with close to 200 science publications and dozens of technical

Figure B.12 The Atacama Pathfinder Experiment (APEX) on its Llano de Chajnantor site. Credit: ESO, CC BY 4.0.

articles. Research has included observations of Galactic molecular clouds and protoplanetary disks, and observations of external galaxies including the Magellanic Clouds.

Atacama Pathfinder Experiment – In a joint venture between MPIfR, OSO, and ESO, a copy of the 12 m diameter ALMA antenna was installed on the Llano de Chajnantor, near ALMA, shown in Figure B.12. The Atacama Pathfinder Experiment (APEX) began operations in 2004. Over 750 scientific articles have resulted from APEX observations, exploring a wide variety of topics. Many of these observations point to further studies with ALMA, but the telescope has proven to be much more than an "ALMA Pathfinder."

Figure B.13 The Large Millimeter Telescope (LMT) at an elevation of 4,600 m on Volcán Sierra Negro, the highest mountain in Mexico. Courtesy of Peter Schloerb, reproduced by permission.

Giant Millimeter Telescopes

Large Millimeter Telescope – A joint project between the United States and Mexico produced the world's largest millimeter/submillimeter wavelength telescope, the Large Millimeter Telescope (LMT), also known as the Gran Telescopio Millimétrico Alfonso Serrano. The partners are the Instituto Nacional de Astrofísica, Óptica y Electrónica, and the University of Massachusetts, Amherst. Figure B.13 shows the telescope in February 2018, shortly after the 50 m diameter reflecting surface was installed. The telescope operates at wavelengths from 0.85 to 4 mm. An image of the disk surrounding the young star ε Eridani was obtained as an early science result using the AzTEC 1.1 mm continuum camera.[11] The LMT has discovered numerous dusty star-forming galaxies at high redshift.[12] It is a key element of the EHT.

Green Bank Telescope – The Robert C. Byrd Green Bank Telescope (GBT) began operation in 2001. The GBT, shown in Figure B.14, is the world's largest fully steerable radio telescope. The subreflector and receiver cabin are offset from the telescope's view of the sky, improving its performance. It has an effective collecting area of 100 m diameter and can support observations at frequencies from 0.1 to 115 GHz. The reflecting surface is made up of 2004 panels

Figure B.14 The Robert C. Byrd Green Bank Telescope in the Allegheny Mountains site of the Green Bank Observatory. Credit: GBO/AUI/NSF, CC BY 3.0.

supported by motor-driven actuators. On calm winter nights, the actuators can be driven to set the reflecting surface to an accuracy that allows for observing at 3 mm wavelength. The GBT has a large instrumentation suite with the flexibility to support a wide variety of studies: pulsars, atomic hydrogen emission, interstellar molecules, dust emission, radar imaging of solar system objects, and very long baseline interferometry.

Notes

1 The simultaneous discoveries of the first CO outflow sources are reported in Snell, Loren, and Plambeck (1980) and Rodriguez et al. (1980).

2 For a more complete history of the MWO, see Vanden Bout, Davis, and Loren (2012).

3 See Dame et al. (1987) and Dame and Thaddeus (2022).

4 Paulikas (1976) gives the 1975 annual report for the Aerospace Corp. Millimeter Telescope. See also Dickman, McCutcheon, and Shuter (1979) and Sargent (1977).

5 The construction of the 7 Meter Telescope is described in a report published in *The Bell System Technical Journal* (Chu et al., 1978).

6 Stark et al. (1988).

7 The origins of IRAM are given in *Open Skies* (Kellerman, Bouton, and Brandt, 2020). Also, *Proposition Commune Pour un Observatoire sur Ondes Millimétriques,* February 1975, (NAA-NRAO, MMA, MMA Planning).

8 Phillips (2009).

9 Lis et al. (2002).

10 A description of the Mt. Fuji Submillimeter Telescope has been published by Sekimoto et al. (2000).

11 See Chavez-Dagostrino et al. (2016).

12 See Berman et al. (2022) and references therein.

Appendix C

Lessons Learned

ALMA is by no means the last international science mega-project to be built by astronomers. In radio astronomy alone we have the Next Generation Very Large Array (ngVLA) being developed by NRAO, and the Square Kilometer Array Observatory (SKAO), approved for construction in Australia and South Africa, with operation planned for the late 2020s. At optical wavelengths, we find the Thirty Meter Telescope (TMT), Giant Magellan Telescope (GMT), and Extremely Large Telescope (ELT), all at varying stages of development and construction. It is worth asking if there are any lessons to be learned from the ALMA experience that should be passed along. The authors have asked past ALMA project managers and directors this question. The synthesis that follows comes primarily from the response by Tony Beasley, the ALMA project manager who called for re-baselining the project at the start of his tenure and managed much of the construction. Comments in the list are also based on conversations with and material[1] from others with experience in ALMA management.

Carry on in the face of disappointment. The loss of the 25 Meter Project was a bitter defeat, but it led to the MMA. Subsequently, the long delay in funding the MMA was discouraging and frustrating. But consider the alternative: if the NSF had funded the MMA in the early 1990s, NRAO would now be in possession of a facility hardly worth operating. During the delay, the concept grew to become ALMA, something vastly more powerful.

Recognize that politics will have a role in the funding of big projects. This is simply a consequence of democratic government; the taxpayers are entitled to a voice and the support of their representatives is essential to a project's success.

Do not underestimate the importance of cultural differences. Indifference to cultural differences causes problems that cost time and money to correct, as well as missed opportunities by not recognizing the value that each partner uniquely brings to the collaboration. Goodwill is helpful but not sufficient. Some differences are deeply embedded and cannot be changed. For example, the process of soliciting bids and selecting contractors is very different in Europe, the United States, and Japan, as are the terms and conditions of contracts. The processes can take different amounts of time. Planning is required to take all of this into account. Patience is certainly an asset. Social behaviors, even in the workplace, also differ from culture to culture, and being conscientious of such differences helps to avoid giving offense. Even something seemingly as straightforward as time zones requires extra effort, for example, scheduling worldwide virtual meetings so that the partners take turns attending at three in the morning. Remember that it always sounds worse in writing than you intended it to be. Despite steadily improving technology to connect the world, face-to-face meetings are often necessary. It certainly was the case with ALMA that in-person interactions were better.

Cultural Differences

In February 2003 the Bilateral ALMA Agreement between North America and Europe was signed. The board chair would rotate every two years and Europe would start. In December 2002, I was elected president of ESO Council and thus became the first chairman of the ALMA Board. After two years, Bob Dickman took over, and I was vice-chairman for another year until my term of president of ESO Council ended.

It was a fascinating, but very busy and definitely tumultuous period. What made it special was that Europe and North America were equal partners. The United States was clearly not accustomed to being in that role, but for Europe, being on equal terms with the United States in major scientific undertakings was also unusual. In Europe, submillimeter astronomy was an extension of optical/infrared to longer wavelengths, while in America it was radioastronomy at higher frequencies. Our communities had different backgrounds, but that did not cause the worst headaches. Often practical, but important matters differed: funding cycles, terms of employment, or project management.

Problems took time to solve and too often compromising was seen as losing face. Early on the Board had to choose an ALMA logo. The design (actually adopted) showed some dishes with four stars arranged as the Southern Cross. The American side at first felt it looked too much like ESO's logo, which also contains that constellation. Then there was the location of the Santiago offices, for which the Vitacura premises of ESO were an obvious option, but only acceptable to the United States if the address would be a different street. These, to Europeans, unimportant issues were resolved but took much time and emotional energy.

There were clashes between personalities, sometimes emotions running very high. No individual, institution, or organization was fully in charge. Persons that had worked hard to realize a millimeter array, saw it develop differently from what they had had in mind. Shaping the collaboration with Chile required diplomacy. We had to deal with cost overruns, deciding on antenna designs, and formalizing the collaboration with Japan into a partnership with yet another culture. The latter made us all minority partners.

Incredibly, in the end every problem was overcome, even the apparently unsolvable ones. It was a great experience working with dedicated persons; I am extremely grateful ALMA came to be and to have been part of it.

Pieter van der Kruit
Jacobus C. Kapteyn Distinguished Professor of Astronomy (Emeritus)
Kapteyn Astronomical Institute, University of Groningen
Groningen, The Netherlands

Effective professional personnel management is critical. Large projects like ALMA have diverse workforces. The employees have a large range of skills and abilities. Effective personnel management requires a dedicated, professional human resources team if the employees are to be treated fairly. Poor workforce morale can be deadly to a project, or at least imply lower efficiency and potentially slowing progress. Employees should be held to a high standard of behavior if they are to work together effectively. Bad behavior should not be tolerated, even in geniuses.

Expect deliveries to be late. It is inevitable that some, perhaps most, deliveries will be late. Contingency plans should be made with this in mind.

Invest heavily in project management, project engineering, and system engineering. The ALMA project discovered rather late that it needed far more systems engineering than had been planned. NRAO had in past projects left systems engineering to a few talented individuals who mainly conversed with one another and produced modest amounts of documentation. This was inadequate for ALMA. ESO had experience in large project management with the VLT, where they relied on massive amounts of documentation to tie the elements of the project together. Reconciling the two approaches took time. Robust document management software with a top-notch search feature greatly facilitates systems engineering.

Remember that there is no substitute for experience – seek knowledge elsewhere. Construction of a complex facility in a high-elevation remote location will encounter unusual situations and questions. It is a mistake to rely on your own staff to solve every knotty problem. The world is full of experts and there is sure to be one with experience in the most arcane area you encounter. Their advice is worth the cost. ALMA learned high-altitude safety standards from the mining industry and high-altitude medicine from Dr. John B. West, a world expert on the subject.

Make decisions promptly, focusing on the big issues and watching risks. Delay leads to increased costs. It is a mistake to put off a decision until every detail has been examined.

Central control of budget and decision-making is critical. ALMA has a flawed management structure. The central office in Chile responsible for ALMA construction and operations is not a legal entity. All purchases and contracts are made by the ALMA Executives: ESO, NRAO, and NAOJ. It required a director and project manager, who were not only capable directors and managers but possessed the diplomatic skills to secure the support of the Executives for their plans and needs. The structure is cumbersome rather than efficient. That it worked at all, and it did, is a testament to the skill, hard work, and goodwill of those involved.

Accept that some decisions will be made politically. No project the size of ALMA can avoid politics. National interests will be asserted and force some decisions. Learn to live with it.

Recognize that the information volume is too high for good external reviews and reporting. No group of experts, however experienced they may be, can fly in for a three-day project review and grasp all that needs to be understood. High-level reviews are essential to maintain commitments to funding, and as such are unavoidable. One can only hope that a review will focus on the high points and issue an endorsement of the project. Accept that detailed advice might miss the mark.

Funding by the partners should be synchronized. Funding for project construction or operation should arrive from all the partners at the same time. Asynchronous funding requires work arounds that can be difficult to achieve. The delay in the availability of funds from the Japanese government greatly complicated the entrance of Japan into ALMA. It was the goodwill and persistence of all those involved that led to success, but at a significant price in time and effort.

Always keep in mind that, in the end, it's all about the science! There will be many occasions – dealing with difficult management, administration, human resources, procurement, and funding issues – that will be extremely frustrating and demoralizing. One of senior management's primary responsibilities is to always exhibit positive leadership to the staff no matter the circumstances, reminding them of the end game, because, *"It is all about the science!"*

Comfort Zones

The management of big science projects presents substantial scientific, engineering, informational, and human resource challenges. When multiple institutions are involved, these challenges become even more complex; when the institutions are from various countries and regions of the world, the challenges can be overwhelming. Like people, organizations have their own personalities, habits, and idiosyncrasies. Those participating in ALMA were no exception. NAOJ is a governmental organization, ESO a treaty-based entity, and AUI a private corporation. None of the Executives shared any common legal basis. These very different legal regimes, in turn, gave rise to very different policies and procedures relating to contracting, human resources, scientific staff, and public outreach.

A successful ALMA project required people to work at a level outside their comfort limits and to effect fundamental changes within their organization. To this end, ALMA was most fortunate. While ALMA certainly did not lack for its share of contentious issues, a core of individuals within each ALMA Executive stepped up to develop creative solutions to seemingly insurmountable problems. These individuals developed effective relations with their counterparts, resulting in a reservoir of personal trust and goodwill which would often sustain them during difficult times. Early in ALMA's existence these small cores of managers recognized that the worst place to thrash out issues were high level meetings, such as the ALMA Board, where protecting an organization's parochial interests was part of the meeting dynamics. Prior to such meetings, people would meet informally, with their counterparts in the other Executives, to explore where the real boundary conditions were so that options for solutions to difficult issues could be devised prior to the actual meetings. Backchanneling made meetings more efficient and often avoided what otherwise might have been embarrassing situations. Formalized agreements were achieved that could not have been envisioned as even possible had people not challenged themselves and their respective organizations to break out of their normal comfort zones.

Certainly, having to meet a construction schedule helped to spur people into actions and risks that they might otherwise not have taken. But without the right set of people in the right places, ALMA's eventual success would have been much riskier. In the operations phase of big projects, the need for a core team of creative thinkers willing to go outside of their personal comfort zone and push against their own organization's normal reflexes continues to be essential. Unfortunately, the need for these qualities often goes unrecognized and under-appreciated.

Patrick W. Donahoe
Associated Universities, Inc. (retired)
Annandale, Virginia

Note

1 A memorandum, *Lessons Learned from ALMA,* by E.J. Schreier and J. Webber, September 2010, is an accounting of the reasons behind the cost increases revealed in the re-baselining of the project. It can be found at NAA-PVB, ALMA, ALMA: The Story of a Science Mega-Project. https://science.nrao.edu/about/publications/alma.

Glossary

Acronyms Used in the Text

AAS	American Astronomical Society
ACA	ALMA Compact Array (Morita Array)
ACAST	NSF Advisory Committee on Astronomical Sciences
ACC	ALMA Coordinating Committee
ACR	ALMA Cost Review
ADC	Analogue to Digital Converter
ADCR	ALMA Delta Cost Review
ADMiT	ALMA Data Mining Toolkit
AEC	ALMA Executive Committee
AEG	Antenna Evaluation Group
AEM	Alcatel Space France, European Industrial Engineering S.r.L., MAN consortium
AIPS	Astronomical Imaging Processing System
AIV	Assembly, Integration, and Verification
ALG	ALMA Liaison Group
ALMA	Atacama Large Millimeter/submillimeter Array
AMAC	ALMA Management Advisory Committee
ANID	Agencia Nacional de Investigación y Desarrollo [Chile]
AOS	Array Operations Site
APEX	Atacama Pathfinder Experiment
ARC	ALMA Regional Center
ARI-L	Additional Representative Images for Legacy
ASA	ALMA Science Archive
ASAC	ALMA Scientific Advisory Committee
ASIAA	Academia Sinica Institute of Astronomy and Astrophysics [Taiwan]
ASP	Astronomical Society of the Pacific
ASPECS	ALMA Spectroscopic Survey
AST	NSF Division of Astronomical Sciences
ASTE	Atacama Submillimeter Telescope Experiment
ATWG	Antenna Technical Working Group

AU	Astronomical Unit
AUI	Associated Universities Inc.
AURA	Associated Universities for Research in Astronomy
AXAF	Advanced X-ray Astronomy Facility
BEC	Business Evaluation Committee
BIMA	Berkeley-Illinois-Maryland Association
BN	Bienes Nacionales [Chile]
BUS	Back Up Structure
CAC	Contract Award Committee
Caltech	California Institute of Technology
CARMA	Combined Array for Research in Millimeter-wave Astronomy
CARTA	Cube Analysis and Rendering Tool for Astronomy
CASA	Common Astronomy Software Applications
CDL	Central Development Laboratory [NRAO]
CERN	Organisation Europeénne pour la Recherche Nucléaire
CfA	Center for Astrophysics
CFRP	Carbon-Fiber-Reinforced Plastic
CfT	Call for Tender
CNRS	Centre National de la Recherche Scientifique [France]
CoC	ESO Committee of Council
CONICYT	Comisión Nacional de Investigación Científica y Tecnológica [Chile]
CSC	Contract Selection Committee
CSO	Caltech Submillimeter Observatory
CSV	Commissioning and Science Verification
CTIO	Cerro Tololo Inter-American Observatory
DSHARP	Disk Substructures at High Angular Resolution
EA	East Asia [ALMA Partner]
EACC	Extended ALMA Coordinating Committee
EU	European [ALMA Partner]
EHT	Event Horizon Telescope
EIE	European Industrial Engineering Consortium
EIS	Environmental Impact Statement
ELT	Extremely Large Telescope
ESO	European Southern Observatory
ESSCO	Electronic Space Systems Corporation
EVALSO	Enabling Virtual Access to Latin American Southern Observatories
EVLA	Expanded Very Large Array
FC	Finance Committee [ESO]
FCRAO	Five College Radio Astronomy Observatory
FIU	Florida International University
GA	Gaz Atacama
GBT	Green Bank Telescope
GIPSY	Groningen Image Processing System
GISS	Goddard Institute for Space Studies
GMC	Giant Molecular Cloud

GMT	Giant Magellan Telescope
HCRO	Hat Creek Radio Observatory
HEMT	High Electron Mobility Transistor
HIAA	Herzberg Institute of Astronomy and Astrophysics [Canada]
HST	Hubble Space Telescope
IAU	International Astronomical Union
IEEE	Institute of Electrical and Electronics Engineers
IfA	Institute for Astronomy [U. Hawaii]
IPT	Integrated Project Team
IRAM	Institute de Radio Astronomie Millimétrique
IRAS	Infrared Astronomy Survey
IS	International Staff
ISM	InterStellar Medium
ITU-R	International Telecommunication Union Radio Communication Sector
JAO	Joint ALMA Office
JATG	Joint Antenna Technical Group
JCMT	James Clerk Maxwell Telescope
JDG	Joint Development Group
JTET	Joint Technical Evaluation Team
KSAI	Korea Space Science Institute
LCO	Las Campanas Observatory
LMA	Large Millimeter Array
LMSA	Large Millimeter Submillimeter Array
LMT	Large Millimeter Telescope
LPG	Liquified Petroleum Gas
LS	Local Staff
LSA	Large Southern Array
MAPS	Molecules with ALMA at Planet-Forming Scales
MELCO	Mitsubishi Electric Company
MEXT	Ministry of Education, Culture, Sports, Science, and Technology [Japan]
MIRIAD	Multichannel Image Reduction, Image Analysis, and Display
MIT	Massachusetts Institute of Technology
MMA	Millimeter Array
MOU	Memorandum of Understanding
MPIA	Max Planck Intitute für Astronomie
MPIfR	Max Planck Institut für Radioastronomie
MPS	Mathematics and Physical Sciences Division of the NSF
MRE	Major Research Equipment
MREFC	Major Research Equipment and Facilities Construction
MWL	Millimeter Wave Laboratory [U. Chile]
MWO	Millimeter Wave Observatory
NA	North America [ALMA Partner]
NAOJ	National Astronomical Observatory of Japan
NAFTA	North American Free Trade Act
NAPRA	North American Program in Radio Astronomy

NASA	National Aeronautics and Space Administration
NCMR	NSF Cost/Management Review
NEON	National Ecological Observatory Network
NFRA	Netherlands Foundation for Radio Astronomy
ngVLA	Next Generation Very Large Array
NINS	National Institutes of Natural Science [Japan]
NMA	Nobeyama Millimeter Array
NOAO	National Optical Astronomy Observatory
NOEMA	NOthern Extended Millimetre Array
NRA	National Rifle Association
NRAO	National Radio Astronomy Observatory
NRC	National Research Council [Canada]
NRO	Nobeyama Radio Observatory
NSB	National Science Board
NSF	National Science Foundation
NWO	Netherlands Research Council
OLPA	Office of Legislative and Public Affairs [NSF]
OSF	Operations Support Facility
OSO	Onsala Space Observatory
OVRO	Owens Valley Radio Observatory
OWL	OverWhelmingly Large telescope
QUARRY	QUabbin ARRaY receiver
PAI	Project Acceptance In-house
PAS	Provisional Acceptance On-Site
PILS	Protostellar Interferometric Line SurveyProtostellar Interferometric Line Survey
PRT	Proposal Receipt Team
PWV	Precipitable Water Vapor
RCL	Radioastronomía Chajnantor Limitada
REBELS	Reionization-Era Bright Line Emission Survey
RFP	Request for Proposals
ROW	Right of Way
SAC	Science Advisory Committee
SAO	Smithsonian Astrophysical Observatory
SCO	Santiago Central Office
SCUBA	Submillimetre Common-User Bolometer Array
SEQUOIA	Second QUabbin Optical Imaging Array
SEST	Sweden-ESO Submillimetre Telescope
SETI	Search for ExtraTerrestrial Intelligence
SIRTF	Space Infrared Telescope Facility (Spitzer)
SIS	Super Conductor-Insulator-Superconductor
SKAO	Square Kilometer Array Observatory
SMA	SubMillimeter Array
SMT	SubMillimeter Telescope [U. Arizona]
SOKENDAI	Graduate University for Advanced Studies

SONG	Survey of Normal Galaxies
STC	ESO's Science and Technology Committee
SUBTEL	Telecommunications Subsecretary of Chile
SUNY	State University of New York
TAC	Technical Advisory Committee
TMT	Thirty Meter Telescope
UCh	Universidad de Chile
UK	United Kingdom
US	United States
VA	Vertex Antennentechnik
VLA	Very Large Array
VLBA	Very Long Baseline Array
VLBI	Very Long Baseline Interferometry
VLT	Very Large Telescope
WBS	Work Breakdown Structure
WSU	Wideband Sensitivity Upgrade

Abbreviations for Journals

A&A	Astronomy and Astrophysics
AJ	Astronomical Journal
ApJ	Astrophysical Journal
ApJL	Astrophysical Journal (Letters)
ApJS	Astrophysical Journal Supplement
ASPC	Astronomical Society of the Pacific Conference (Series)
ASR	Advances in Space Research
BAAS	Bulletin of the American Astronomical Society
JAHH	Journal of Astronomical History and Heritage
JRASC	Journal of the Royal Astronomical Society of Canada
LNP	Lecture Notes in Physics
MNRAS	Monthly Notices of the Royal Astronomical Society
PASJ	Publications of the Astronomical Society of Japan
PASP	Publications of the Astronomical Society of the Pacific
PhRvL	Physical Review Letters
PJAB	Proceedings of the Japan Academy, Series B

Citation Abbreviations for NRAO/AUI Archives Material

NAA	NRAO/AUI Archives
NAA-AUI	NRAO/AUI Archives, Records of Associated Universities, Inc.
NAA-NRAO	NRAO/AUI Archives, Records of the National Radio Astronomy Observatory
NAA-PVB	NAA/NRAO, Papers of Paul A. Vanden Bout
NAA-RLB	NAA/NRAO, Papers of Robert L. Brown

References

Abe, K., Tsutsumi, J., and Hiyama, T. 2014, ACA Correlator System: Supercomputer System Developed for ALMA Project, *FUJITSU Sci. Tech. J.*, **50**, 35

Akiyama, K., et al. 2019, First M87 Event Horizon Telescope Results. I. The Shadow of the Supermassive Black Hole, *ApJL*, **875**, L1

Akiyama, K., et al. 2021, First M87 Event Horizon Telescope Results. VIII. Magnetic Field Structure Near the Event Horizon, *ApJL*, **910**, L13

ALMA Partnership, et al. 2015, The 2014 ALMA Long Baseline Campaign: First Results from High Angular Resolution Observations toward the HL Tau Region, *ApJL*, **808**, L3

Ando, R., et al. 2017, Diverse Nuclear Star-Forming Activities in the Heart of NGC 253 Resolved with 10-pc-Scale ALMA Images, *ApJ*, **849**, 81

Andrews, S. M., et al. 2018, The Disk Substructures High Resolution Project (DSHARP). I. Motivation, Sample, Calibration, and Overview, *ApJL*, **869**, L41

Baars, J. W. M., et al. 2007, Near-Field Radio Holography of Large Reflector Antennas, *IEEE Antennas Propag. Mag.*, **49**, 24

Bae, J., et al. 2022, Molecules with ALMA at Planet-forming Scales (MAPS): A Circumplanetary Disk Candidate in Molecular-line Emission in the AS209 Disk, *ApJL*, **934**, L20

Bastian, T. S., et al. 2017, A First Comparison of Millimeter Continuum and Mg II Ultraviolet Line Emission from the Solar Chromosphere, *ApJL*, **845**, L19

Baudry, A., et al. 2012, Performance Highlights of the ALMA Correlators. In SPIE Proc. **8452**, *Millimeter, Submillimeter, and Far-Infrared Detectors and Instrumentation for Astronomy VI*, ed. W. S. Holland (Bellingham, WA: SPIE), 8452-17

Berman, D. A., et al. 2022, PASSAGES: The Large Millimeter Telescope and ALMA Observations of Extremely Luminous High Redshift Galaxies Identified by the Planck, *MNRAS*, **515**, 3911

Bok, B. and Reilly, E. F. 1947, Small Dark Nebulae, *ApJ*, **105**, 25

Booth, R. S. 1994, A Southern Hemisphere Millimeter Array. In ASPC **59**, *Astronomy with Millimeter and Submillimeter Wave Interferometry, IAU Colloquium 140*, ed. M. Ishiguro and W. J. Welch (San Francisco: ASP), 413

Booth, R. S. 1997, European Plans for a Millimetre Array. In IAU Symposium No. **170**, *CO: Twenty-Five Years of Millimeter-Wave Spectroscopy*, ed. W. B. Latter et al. (Dordrecht: Kluwer), 231

Brogan, C., et al. 2019, A Science-Driven Vision for ALMA in the 2030s, *BAAS*, **51**, 7

Brown, R. L. 1997, New Instruments, New Science: Future Opportunities. In IAU Symposium No. **170**, *CO: Twenty-Five Years of Millimeter-Wave Spectroscopy*, ed. W. B. Latter et al. (Dordrecht: Kluwer), 247

Brown, R. L., and Vanden Bout, P. A. 1991, CO Emission at z=2.2867 in the Galaxy IRAS F10214+4724, *AJ*, **102**, 1956.

Carpenter, J. M., et al. 2022, Update on the Systematics in the ALMA Proposal Review Process after Cycle 8, *PASP*, **134**, 1

Chavez-Dagostino, M., et al. 2016, Early Science with the Large Millimetre Telescope: Deep LMT/AzTEC Millimetre Observations of ε Eridani and Its Surroundings, *MNRAS*, **462**, 2285

Chu, T. S., et al. 1978, The Crawford Hill 7-Meter Millimeter Antenna, *Bell Syst. Tech. J.*, **57**, 1257

Crease, R. P. 1990, Millimeter Scientists Push for New Telescope, *Science*, **249**, 1504

Dame, T. M., et al. 1987, A Composite CO Survey of the Entire Milky Way, *ApJ*, **332**, 706

Dame, T. M., and Thaddeus, P. 2022, A CO Survey of the Entire Northern Sky, *ApJS*, **262**, 5

David, P. 1983, Millimetre Wave Astronomy: US Could Be Left Behind, *Nature*, **303**, 7

Delannoy, J., Lacroix, J., and Blum, E. J. 1973, An 8-mm Interferometer for Solar Radio Astronomy at Bordeaux, France, *Proc. IEEE*, **61**, 1282

Dickman, R. L. 1978, The Ratio of Carbon Monoxide to Molecular Hydrogen in Interstellar Dark Clouds, *ApJS*, **37**, 407

Dickman, R. L., McCutcheon, W. H., and Shuter, W. L. H. 1979, Carbon Monoxide Isotope Fractionation in the Dust Cloud L134, *ApJ*, **234**, 100

Donovan Meyer, J., et al. 2022, Analysis of the ALMA Cycle-8 Distributed Review Process, *BAAS*, **54**, 1

Downes, D. 1994, New Directions for Millimeter Astronomy in the 21st Century. In *Frontiers of Space and Ground-Based Astronomy 1994*, ed. W. Wamsteker, M. S. Longair, and Y. Kondo (Dordrecht: Kluwer), 133

Drake, F., and Sobel, D. 1994, *Is Anyone Out There? The Scientific Search for Extraterrestrial Intelligence* (New York: Delecourt)

Escoffier, R. P., et al. 2007, The ALMA Correlator, *A&A*, **462**, 801

Fudamoto, Y., et al. 2021, Normal, Dust-obscured Galaxies in the Epoch of Reionization, *Nature*, **597**, 489

Gordon, M. A. 2005, *Recollections of "Tucson Operations": The Millimeter-Wave Observatory of the National Radio Astronomy Observatory* (Dordrecht: Springer)

Herrera, C. N., et al. 2012, ALMA CO and VLT/SINFONI H_2 Observations of the Antennae Overlap Region: Mass and Energy Dissipation, *A&A*, **538**, L9

Herzberg, G. 1988, Historical Remarks on the Discovery of Interstellar Molecules, *JRASC*, **82**, 115

Hills, R. E., et al. 1972, Interferometric Positions of the Water-Vapor Emission Sources in HII Regions, *ApJL*, **175**, L59

Hills, R. E., et al. 1973, The Hat Creek Millimeter Wave Interferometer, *Proc. IEEE*, **61**, 127

Iguchi, S., et al. 2009, The ALMA Compact Array (ACA), *PASJ*, **61**, 1

Ishiguro, M., et al. 1994, The Large Millimeter Array. In ASPC 59, *Astronomy with Millimeter and Submillimeter Wave Interferometry, IAU Colloquium 140*, ed. M. Ishiguro and W. J. Welch (San Francisco: ASP), 405

Ishiguro, M. 1997, LMSA: Japanese Plans for a Large Millimeter and Submillimeter Array. In IAU Symposium No. **170**, *CO: Twenty-Five Years of Millimeter-Wave Spectroscopy*, ed. W. B. Latter et al. (Dordrecht: Kluwer), 239

Ishiguro, M., and the LMSA Working Group. 1998, Japanese Large Millimeter and Submillimeter Array. In SPIE Proc. **3357**, *Advanced Technology MMW, Radio and Terahertz Telescopes*, ed. T. G. Phillips (Bellingham, WA: SPIE), 244

Ishiguro, M. 2009, *ALMA Radio Telescope* (Tokyo: Chikumashobo) [in Japanese]

Ishiguro, M., Chiba, K., and Sakamoto, S. 2022, From Nobeyama Radio Observatory to the International Project ALMA: Evolution of Millimeter and Submillimeter Wave Astronomy in Japan, *PJAB*, **98**, 439

Iwai, K., et al. 2017, ALMA Discovery of Solar Umbral Brightness Enhancement at λ=3 mm, *ApJL*, **841**, L20

Janssen, M., et al. 2021, Event Horizon Telescope Observations of the Jet Launching and Collimation in Centaurus A, *Nature Astronomy*, **5**, 1017

Jorgensen, J. K., et al. 2016, The ALMA Protostellar Interferometric Line Survey (PILS). First Results from an Unbiased Submillimeter Wavelength Line Survey of the Class 0 Protostellar Binary IRAS 16293-2422 with ALMA, *A&A*, **595**, A117

Kellermann, K. I., Bouton, E. N., and Brandt, S. S. 2020, *Open Skies: The National Radio Astronomy Observatory and Its Impact on US Radio Astronomy* (Cham: Springer)

Kellermann, K. I., and Bouton, E. N. 2023, *Star Noise: Discovering the Radio Universe* (Cambridge: Cambridge University Press)

Koda, J., et al. 2009, Dynamically Driven Evolution of the Interstellar Medium in M51, *ApJL*, **700**, L132

Kono, K., et al. 1995, Preliminary Result of Site Testing in Northern Chile with a Portable 220 GHz Radiometer, *NRO Technical Report*, No. 42

Kwon, W., Looney, L. W., and Mundy, L. G. 2011, Resolving the Circumstellar Disk in HL Tauri at Millimeter Wavelengths, *ApJ*, **741**, 3

Latter, W. B., et al., eds. 1997, IAU Symposium No. 170, *CO: Twenty-Five Years of Millimeter-Wave Spectroscopy* (Dordrecht: Kluwer)

Ligterink, N. F. W., et al. 2017, The ALMA-PILS Survey: Detection of CH3NCO Towards the Low-Mass Protostar IRAS 16293-2422 and Laboratory Constraints on its Formation, *MNRAS*, **469**, 2219

Lis, D. C., et al. 2002, The Role of Outflows and C Shocks in the Strong Deuteration of L1689, *ApJ*, **571**, L5

Loukitcheva, M. A. 2019, First Solar Observations with ALMA, *ASR*, **63**, 1396

Loukitcheva, M. A., et al. 2017, Solar ALMA Observations: Constraining the Chromosphere above Sunspots, *ApJ*, **850**, 35

Lynds, B. 1962, Catalog of Dark Nebulae, *ApJS*, **7**, 1

Madsen, K. 2012, *The Jewel on the Mountaintop* (Berlin: Wiley CVH)

Mangum, J. G., et al. 2006, Evaluation of the ALMA Prototype Antennas, *PASP*, **118**, 1257

Martín-Doménech, R., et al. 2017, Detection of Methyl Isocyanate (CH3NCO) in a Solar-Type Protostar, *MNRAS*, **469**, 2230

McGuire, B. A. 2022, 2021 Census of Interstellar, Circumstellar, Extragalactic, Protoplanetary Disk, and Exoplanetary Molecules, *ApJS*, **259**, 30

Otárola, A., Delgado, G., and Bååth, L. 1995, Site Survey for a Large Southern Array. In *Science with Large Millimeter Arrays*, ed. P. A. Shaver (Dordrecht: Springer), 358

Paulikas, G. A. 1976, The Aerospace Corp., El Segunda, California. Observatory Report, *BAAS*, **8**, 1

Peek, K. 2017, Long Live Hubble, *Sci. Amer.*, **316**, 80

Phillips, T. G. 2009, Development of the Submillimeter Band. In ASPC **417**, *Submillimeter Astrophysics and Technology: A Symposium Honoring Thomas G. Phillips*, ed. D. C. Lis et al. (San Francisco: ASP), 37

Plambeck, R. L. 2006, The Legacy of the BIMA Millimeter Array. In ASPC **356**, *Revealing the Molecular Universe: One Antenna Is Never Enough*, ed. D. C. Backer, J. L. Turner, and J. M. Moran (San Francisco: ASP), 3

Pun, M., et al. 2018, Effects on Cognitive Functioning of Acute, Subacute and Repeated Exposures to High Altitude, *Front. Physiol.*, **9**, 1131

Radford, S. J. E., et al. 1996, Resolution of the Discrepancy in the CO(3-2) Flux of IRAS F10214+4724, *AJ*, **111**, 1021

Rizzo, F., et al. 2020, A Dynamically Cold Disk Galaxy in the Early Universe, *Nature*, **584**, 201

Rodriguez, L. F., et al. 1980, Anisotropic Mass Outflow in Cepheus A, *ApJ*, **240**, L149

Sakamoto, S., 2001, Comparison of the Pampa La Bola and Llano de Chajnantor Sites in Northern Chile. In ASPC, **266**, *Astronomical Site Evaluation in the Optical and Radio Range*, ed. J. Vernin, Z. Benkahldun, and C. Muñoz-Tuñón (San Francisco: ASP), 440

Sargent, A. I. 1977, Molecular Clouds and Star Formation. I. Observations of the Cepheus OB3 Molecular Cloud, *ApJ*, **218**, 736

Sargent, A. I., and Beckwith, S. 1987, Kinematics of the Circumstellar Gas of HL Tauri & R Monocerotis, *ApJ*, **323**, 294

Schnee, S. L., et al. 2014, The Human Pipeline: Distributed Data Reduction for ALMA. In *Proc. SPIE*, **9149**, *Observatory Operations: Strategies, Processes, and Systems V*, ed. A. B. Peck, C. R. Benn, and R. L. Seaman (Bellingham, WA: SPIE), 9149–0Z

Scoville, N. Z., et al. 1986, Millimeter Interferometry of the Molecular Gas in Arp 220, *ApJL*, **311**, L47

Sekimoto, Y., et al. 2000, The Mt. Fuji Submillimeter-wave Telescope. In *Proc. SPIE* **4015**, *Radio Telescopes*, ed. H. Butcher (Bellingham, WA: SPIE), 185

Shaver, P. A., 1986, *Science with Large Millimetre Arrays* (Berlin: Springer)

Shillue, B., et al. 2012, The ALMA Photonic Local Oscillator System. In *Proc. SPIE* **8452**, *Millimeter, Submillimeter, and Far-Infrared Detectors and Instrumentation for Astronomy VI*, ed. W. S. Holland (Bellingham, WA: SPIE), 8452–16

Shimojo, M., et al. 2017, The First ALMA Observation of a Solar Plasmoid Ejection from an X-Ray Bright Point, *ApJL*, **841**, L5

Snell, R. L., Loren, R. B., and Plambeck, R. L. 1980, Observations of CO in L1551: Evidence for Stellar Wind Driven Shocks, *ApJ*, **239**, L17

Snyder, L. E., et al. 1969, Microwave Detection of Interstellar Formaldehyde, *PhRvL*, **22**, 679

Solomon, P. M., Downes, D., and Radford, S. J. E. 1992, Warm Molecular Gas in the Primeval Galaxy IRAS 10214 +4724, *ApJL*, **398**, L29

Solomon, P. M., and Wickramasinghe, N. C. 1969, Molecular and Solid Hydrogen in Dense Interstellar Clouds, *ApJ*, **158**, 449

Stark, A., et al. 1988, The Bell Laboratories CO Survey. In LNP **315**, *Molecular Clouds in the Milky Way and External Galaxies,* ed. R. Dickman, R. Snell, and J. Young (Heidelberg: Springer), 303

Townes, C. H. 1957, Microwave and Radio-Frequency Resonance Lines of Interest to Radio Astronomy. In IAU Symposium No. 4, *Radio Astronomy,* ed. H. C. van de Hulst (Dordrecht: Reidel), 92

Townes, C. H. 2006, The Discovery of Interstellar Water Vapor and Ammonia at the Hat Creek Radio Observatory. In ASPC **356**, *Revealing the Molecular Universe: One Antenna Is Never Enough,* ed. D. C. Backer, J. L. Turner, and J. M. Moran (San Francisco: ASP), 81

Tsukui, T., and Iguchi, S. 2021, Spiral Morphology in an Intensely Star-Forming Disk Galaxy More Than 12 Billion Years Ago, *Science,* **372**, 1201

Ukita, N., et al. 2004, Design and Performance of the ALMA-J Prototype Antenna. In Proc. SPIE **5489**, *Ground-based Telescopes,* ed. J. M. Oschman, Jr. (Bellingham WA: SPIE), 1089

Vanden Bout, P. A. 2005, Origins of the ALMA Project in the Scientific Visions of the North American, European, and Japanese Astronomical Communities. In ESA **SP-577**, *The Dusty and Molecular Universe,* ed. A. Wilson (Noordwijk: ESA), 23

Vanden Bout, P. A., Davis, J. H., and Loren, R. B. 2012, The University of Texas Millimeter Wave Observatory, *JAHH,* **15**, 232

Waldrop, M. M. 1983, Astronomers Ponder a Catch-22, *Science,* **220**, 698

Walter, F., et al. 2016, ALMA Spectroscopic Survey in the Hubble Deep Field: Survey Description, *ApJ,* **833**, 67

Wedemeyer, S., et al. 2020, The Sun at Millimeter Wavelengths. I. Introduction to ALMA Band 3 Observations, *A&A,* **635**, A71

Weinreb, S., et al. 1963, Radio Observations of OH in the Interstellar Medium, *Nature,* **200**, 829

Welch, W. J. 1996, The Berkeley-Illinois-Maryland Association Millimeter Array, *PASP,* **108**, 93

Wilson, R. W., Jefferts, K. B., and Penzias, A. A. 1970, Carbon Monoxide in the Orion Nebula, *ApJ,* **161**, L43

Wilson, R. W. 2008, Discovering CO and Other Interstellar Molecules with the NRAO 36 Foot Antenna. In ASPC **395**, *Frontiers of Astrophysics: A Celebration of NRAO's 50th Anniversary,* ed. A. H. Bridle, J. J. Condon, and G. C. Hunt (San Francisco: ASP), 183

Wilson, R. W. 2015, The Discovery of Interstellar CO. In *26th International Symposium On Space Terahertz Technology,* W1-1 [www.nrao.edu/meetings/isstt/papers/2015/2015000034.pdf]

Woody, D., Vail, D., and Schall, W. 1994, Design, Construction, and Performance of the Leighton 10.4-m- diameter Radio Telescopes, *Proc. IEEE,* **82**, 673

Woody, D., et al. 2004, CARMA: A New Heterogeneous Millimeter Wave Interferometer. In Proc. SPIE **5498**, *Millimeter and Submillimeter Detectors for Astronomy II,* ed. J. Zmuidzinas, W. S. Holland, and S. Withington (Bellingham WA: SPIE), 30

Wootten, A., and Thompson, A. R. 2009, The Atacama Large Millimeter/submillimeter Array, *Proc. IEEE,* **97**, 1463

Yamane, K. 2017, *Creators of the Super Telescope "ALMA"* (Konsarutingu: Nikkei BP) [in Japanese]

Index

Printed in the United States
by Baker & Taylor Publisher Services